Praise for Mark O'Connell's

To Be a Machine

"O'Connell, a columnist for *Slate,* is a charming, funny tour guide. Writing on transhumanism often gets swept away by the inherent drama of its adherents' promises, but O'Connell's eye for small human details... keeps the narrative grounded in a way that rigorous scientific debunking wouldn't." —*Vice*

"O'Connell unleashes his prodigious researching and writing skills on what could be your future." —*The Philadelphia Inquirer*

"*To Be a Machine* is an attempt to understand the transhumanist movement on its own terms.... It's O'Connell's lack of stridency, as well as his often splendid writing, that makes him such a companionable guide." —*The Guardian*

"O'Connell has devised an indispensable GPS for negotiating today's tomorrow-land." —*The Free Lance-Star* (Fredericksburg, VA)

"[O'Connell] reveals a bounty of beguiling ingenuity and genuine absurdity, eliciting laughs and empathy, because we are our most human while trying to become something more than human." —*Playboy*

"[O'Connell] dissects the practices and beliefs of transhumanism with extraordinary exuberance and wit.... *To Be a Machine* is sometimes hilarious (triggering several bursts of uncontrollable giggles while I read it on the Tube), but even as O'Connell mocks the more absurd manifestations of transhumanism, he shows sympathy and understanding for its adherents." —Clive Cookson, *Financial Times*

"O'Connell, a journalist, makes his own prejudices clear: 'I am not now, nor have I ever been, a transhumanist,' he writes. However, this does not stop him from thoughtfully surveying the movement."
—*Science*

"O'Connell's book is skeptical but not cynical, and it functions as a witty overview of transhumanism." —*The Ringer*

"O'Connell invokes the twin specters of death and childbearing in an attempt to make sense of his subject—but he also manages to be staggeringly funny." —*New Scientist*

"O'Connell's sensibility—his humanity, if you will—and his subject matter are a match made in heaven. It's an absolutely wonderful book." —*The Millions*

"Troubling and humorous, this is one of my current give-it-to-everyone books—I buy six copies at a time. Did you know our future belongs to a few asocial geeks for whom being human has always been a problem? Now they can solve it."
—Jeanette Winterson, *Vulture*

"Flat-out fascinating.... Sharply written and thought-provoking."
—*The Maine Edge*

"O'Connell writes with an intellectual curiosity that makes his esoteric subject matter accessible to lay readers.... A stimulating overview of modern scientific realities once thought to be the exclusive purview of science fiction." —*Publishers Weekly*

"By exposing the ludicrous yet terrifyingly serious ideologies behind transhumanism, *To Be a Machine* is an important book, as well as a seriously funny one." —*The Sunday Times* (London)

MARK O'CONNELL

To Be a Machine

Mark O'Connell is *Slate*'s books columnist, a staff writer at *The Millions*, and a regular contributor to *The New Yorker*'s "Page-Turner" blog. His work has been published in *The New York Times Magazine*, *The New York Times Book Review*, *The Observer*, and *The Independent*. He lives in Dublin.

www.mark-oconnell.com

MARK O'CONNELL

To Be a Machine

Adventures Among Cyborgs, Utopians, Hackers, and the Futurists Solving the Modest Problem of Death

ANCHOR BOOKS
A Division of Penguin Random House LLC
New York

FIRST ANCHOR BOOKS EDITION, JANUARY 2018

 Published in the United States by Anchor Books, a division of Penguin Random House LLC, New York, and distributed in Canada by Random House of Canada, a division of Penguin Random House Canada Limited, Toronto. Originally published in hardcover in the United States by Doubleday, a division of Penguin Random House LLC, New York, in 2017.

The Library of Congress has cataloged the Doubleday edition as follows:
Names: O'Connell, Mark, [date] author.
Title: To be a machine : adventures among cyborgs, utopians, hackers, and the futurists solving the modest problem of death / Mark O'Connell.
Description: First edition. | New York: Doubleday, 2017.
Identifiers: LCCN 2016021725
Subjects: LCSH: Humanism. | Medical technology–Social aspects. | Technological innovations–Social aspects. | Prosthesis–Social aspects. | BISAC: COMPUTERS / Intelligence (AI) & Semantics. | SCIENCE / Biotechnology. | COMPUTERS / Social Aspects / Human-Computer Interaction.
Classification: LCC B821.O365 2016 | DDC 306.4/61–dc23
LC record available at https://lccn.loc.gov/2016021725

Anchor Books Trade Paperback ISBN: 978-1-101-91159-4
eBook ISBN: 978-0-385-54042-1

Author photograph © Richard Gilligan
Book design by Michael Collica

www.anchorbooks.com

146122990

For Amy and Mike, for everything

This is the whole point of technology.
It creates an appetite for immortality on the one hand.
It threatens universal extinction on the other.
Technology is lust removed from nature.

—Don DeLillo, *White Noise*

Contents

To Be a Machine

System Crash

ALL STORIES BEGIN in our endings: we invent them because we die. As long as we have been telling stories, we have been telling them about the desire to escape our human bodies, to become something other than the animals we are. In our oldest written narrative, we find the Sumerian king Gilgamesh, who, distraught by the death of a friend and unwilling to accept that the same fate lies in store for him, travels to the far edge of the world in search of a cure for mortality. Long story short: no dice. Later, we find Achilles' mother dipping him in the Styx in an effort to render him invulnerable. This, too, famously, does not pan out.

See also: Daedalus, improvised wings.

See also: Prometheus, stolen divine fire.

We exist, we humans, in the wreckage of an imagined splendor. It was not supposed to be this way: we weren't supposed to be weak, to be ashamed, to suffer, to die. We have always had higher notions of ourselves. The whole setup–garden, serpent, fruit, banishment–was a fatal error, a system crash. We came to be what we are by way of a Fall, a retribution. This, at least, is one version of the story: the Christian story, the Western story. The point of which, on some level, is to explain ourselves to ourselves, to account for why it's such a raw deal, this unnatural nature of ours.

"A man," wrote Emerson, "is a god in ruins."

Religion, more or less, arises out of this divine wreckage. And science, too—religion's estranged half sibling—addresses itself to such animal dissatisfactions. In *The Human Condition*, writing in the wake of the Soviet launch of the first space satellite, Hannah Arendt reflected on the resulting sense of euphoria about escaping what one newspaper report called "men's imprisonment to the earth." This same yearning for escape, she wrote, manifested itself in the attempt to create superior humans from laboratory manipulations of germ plasm, to extend natural life spans far beyond their current limits. "This future man," she wrote, "whom the scientists tell us they will produce in no more than a hundred years, seems to be possessed by a rebellion against human existence as it has been given, a free gift from nowhere (secularly speaking), which he wishes to exchange, as it were, for something he has made himself."

A rebellion against human existence as it has been given: this is as good a way as any of attempting to encapsulate what follows, to characterize what motivates the people I came to know in the writing of this book. These people, by and large, identify with a movement known as transhumanism, a movement predicated on the conviction that we can and should use technology to control the future evolution of our species. It is their belief that we can and should eradicate aging as a cause of death; that we can and should use technology to augment our bodies and our minds; that we can and should merge with machines, remaking ourselves, finally, in the image of our own higher ideals. They wish to exchange the gift, these people, for something better, something man-made. Will it pan out? That remains to be seen.

I am not a transhumanist. That much is probably apparent, even at this early stage of the proceedings. But my fascination with the movement, with its ideas and its aims, arises out of a basic sympathy with its premise: that human existence, as it has been given, is a suboptimal system.

In an abstract sort of way, this is something I had always believed to be the case, but in the immediate aftermath of the birth of my son, I came to feel it on a visceral level. The first time I held him, three years ago now, I was overcome by a sense of the fragility of his little body—a body that had just emerged, howling and trembling and darkly smeared with blood, out of the trembling body of his mother, from whom many hours of fanatical suffering and exertion had been required to deliver him into the world. *In sorrow thou shalt bring forth children.* I couldn't help but think that there ought to be a better system. I couldn't help but think that, at this late stage, we should be beyond all this.

Here's a thing you should not do as a new father, as you perch uneasily on a leatherette maternity ward chair beside your sleeping infant and his sleeping mother: you should not read a newspaper. I did this, and I regretted it. I sat in the postnatal ward of the National Maternity Hospital in Dublin, turning the pages of *The Irish Times* in gradually mounting horror, browsing through a catalog of human perversity—of massacres and rapes, of cruelties casual and systemic: splintered dispatches from a fallen world—and wondered about the wisdom of bringing a child into this mess, this species. (I seem to remember having a mild head cold at the time; this would not have helped matters.)

Among its many other effects, becoming a parent forces you to think about the nature of the problem—which is, in a lot of ways, the problem of nature. Along with all the other horrors and perversities of the broader human context, the realities of aging and sickness and mortality become suddenly inescapable. Or they did for me, at any rate. And for my wife, too, whose existence was so much more entangled with our son's in those early months, and who said something during that time that I will never forget. "If I had known how much I was going to love him," she said, "I'm not sure I would have had him." The frailty is the thing, the vulnerability. This infirmity, this doubtful convalescence we refer to, for want of a better term, as the human condition. Condition: an illness or other medical problem.

For dust thou art, and unto dust shalt thou return.

In hindsight, it seems like more than mere coincidence that this was the period during which I became obsessed with an idea I'd first encountered close to a decade previously, and which was now beginning to consume my thoughts—the idea that this condition might not be our ineluctable fate. That like nearsightedness or smallpox, it might be set to rights by the intervention of human ingenuity. I was obsessed, that is, for the same reason as I had always been obsessed by the story of the Fall, and the notion of original sin: because it expressed something profoundly true about the deepest strangeness of being human, our inability to accept ourselves, our capacity to believe we might be redeemed of our nature.

Early on in the pursuit of this obsession—a pursuit that had, at that point, yet to lead beyond the Internet into what is fondly referred to as the "real world"—I came upon a strange and provocative text entitled "A Letter to Mother Nature." It was, as its name suggested, a kind of epistolary manifesto addressed to the anthropomorphic figure to whom, for the sake of clarity, the creation and husbandry of the natural world is often attributed. The text, in an initial tone of mild passive aggression, began by thanking Mother Nature for her mostly solid work on the project of humanity thus far, for raising us from simple self-replicating chemicals to trillion-celled mammals with the capacity for self-understanding and empathy. The letter then smoothly transitioned into full *J'accuse* mode, briefly outlining some of the more shoddy workmanship evident in the functioning of *Homo sapiens:* the vulnerability to disease and injury and death, for instance, the ability to function only in highly circumscribed environmental conditions, the limited memory, the notoriously poor impulse control.

The author—addressing Mother Nature as the collective voice of her "ambitious human offspring"—then proposed a total of seven amendments to "the human constitution." We would no longer consent to live under the tyranny of aging and death, but would use the tools of biotechnology to "endow ourselves with enduring vitality and remove our expiration date." We would augment our powers of perception and cognition through technological enhancements of our sense organs and our neural capacities. We would no longer submit to being the products of blind evolution, but would rather "seek complete choice of bodily form and function, refining and augmenting our physical and intellectual abilities beyond those of any human in history." And we would no longer be content to limit our physical, intellectual, and emotional capacities by remaining confined to carbon-based biological forms.

This "Letter to Mother Nature" was the clearest and most provocative statement of transhumanist principles I had encountered, and its epistolary conceit captured something crucial about what made the movement so strange and compelling to me—it was direct, and audacious, and it pushed the project of Enlightenment humanism to such radical extremes that it threatened to obliterate it entirely. There was, I felt, a whiff of madness about the whole enterprise, but it was a madness that revealed something fundamental about what we thought of as reason. The letter was, I learned, the work of a man who went by the thematically consistent name Max More—an Oxford-educated philosopher who turned out to be one of the central figures in the transhumanist movement.

There was, I came to see, no one accepted or canonical version of this movement; but the more I read about it, and the more I came to understand the views of its adherents, the more I understood it as resting on a mechanistic view of human life—a view that human beings were devices, and that it was our duty and our destiny to become better versions of the devices that we were: more efficient, more powerful, more useful.

I wanted to know what it meant to think of yourself, and more broadly your species, in such instrumentalist terms. And I wanted to know more specific things: I wanted to know, for instance, how you might go about becoming a cyborg. I wanted to know how you might upload your mind into a computer or some other hardware, with the aim of existing eternally as code. I wanted to know what it would mean to think of yourself as no more or less than a complex pattern of information, as no more or less than code. I wanted to learn what robots might disclose about our understanding of ourselves and our bodies. I wanted to know how likely artificial intelligence was to redeem or annihilate our species. I wanted to know what it might be like to have faith in technology sufficient to allow a belief in the prospect of your own immortality. I wanted to learn what it meant to be a machine, or to think of yourself as such.

And I did, I assure you, arrive at some answers to these questions along the way; but in investigating what it meant to be a machine, I must tell you that I also wound up substantially more confused than I already was about what it meant to be a human being. More goal-oriented readers should be advised, therefore, that this book is as much an investigation of that confusion as it is an analysis of those learnings.

A broad definition: transhumanism is a liberation movement advocating nothing less than a total emancipation from biology itself. There is another way of seeing this, an equal and opposite interpretation, which is that this apparent liberation would in reality be nothing less than a final and total enslavement to technology. We will be bearing both sides of this dichotomy in mind as we proceed.

For all the extremity of transhumanism's aims—the convergence of technology and flesh, for instance, or the uploading of minds into machines—the above dichotomy seemed to me to express something fundamental about the particular time in which we find ourselves, in which we are regularly called upon to consider how technology is changing everything for the better, to acknowledge the extent to which a particular app or platform or device is making the world a better place. If we have hope for the future—if we think of ourselves as having such a thing as a future—it is predicated in large part on what we might accomplish through our machines. In this sense, transhumanism is an intensification of a tendency already inherent in much of what we think of as mainstream culture, in what we may as well go ahead and call capitalism.

And yet the inescapable fact of this aforementioned moment in history is that we, and these machines of ours, are presiding over a vast project of annihilation, an unprecedented destruction of the world we have come to think of as ours. The planet is, we are told, entering a sixth mass extinction: another Fall, another expulsion. It seems very late in the day, in this dismembered world, to be talking about a future.

One of the things that drew me to this movement, therefore, was the paradoxical force of its anachronism. For all that transhumanism presented itself as resolutely oriented toward a vision of a world to come, it felt to me almost nostalgically evocative of a human past in which radical optimism seemed a viable position to take with respect to the future. In the way it looked forward, transhumanism seemed, somehow, always to be facing backward.

The more I learned about transhumanism, the more I came to see that, for all its apparent extremity and strangeness, it was nonetheless exerting certain formative pressures on the culture of Silicon Valley, and thereby the broader cultural imagination of technology. Transhumanism's influence seemed perceptible in the fanatical dedication of many tech entrepreneurs to the ideal of radical life extension—in the PayPal cofounder and Facebook investor Peter Thiel's funding of various life extension projects, for instance, and in Google's establishment of its biotech subsidiary Calico, aimed at generating solutions to the problem of human aging. And the movement's influence was perceptible, too, in Elon Musk's and Bill Gates's and Stephen Hawking's increasingly vehement warnings about the prospect of our species' annihilation by an artificial superintelligence, not to mention in Google's instatement of Ray Kurzweil, the high priest of the Technological Singularity, as its director of engineering. I saw the imprint of transhumanism in claims like that of Google CEO Eric Schmidt, who suggested that "Eventually, you'll have an implant, where if you just think about a fact, it will tell you the answer." These men—they were men, after all, almost to a man—all spoke of a future in which humans would merge with machines. They spoke, in their various ways, of a posthuman future—a future in which techno-capitalism would survive its own inventors, finding new forms in which to perpetuate itself, fulfill its promise.

Not long after I read Max More's "Letter to Mother Nature," I came across a film on YouTube called *Technocalyps,* a 2006 documentary about transhumanism by a Belgian filmmaker named Frank Theys, one of only a very small number of films I'd been able to find about the movement. Midway through the film, there is a brief sequence in which a young man, fair-haired and bespectacled, dressed entirely in black, stands alone in a room and performs an odd ritual. The scene is dimly lit, and shot on what seems to be a webcam, and so it is difficult to tell exactly where we are. It looks to be a bedroom, though there are computers on a desk in the background, so it could just as easily be an office. These computers, with their beige desktop towers and their squatly cuboid monitors, seem to date the film to around the turn of this present century. Against this background, the young man stands facing us, both arms raised above his head in an oddly hieratic gesture. In a lilting Scandinavian staccato that lends his voice a mechanistic quality, he begins to speak.

"The data, the code, the communications," he says. "Forever, amen."

With these invocations, he moves his arms downward, then outward to either side, before clasping his hands to his chest. He turns about the room, bestowing a gesture of esoteric benediction on the four points of the compass, speaking in each of these positions the hallowed name of a prophet of the computer age: Alan Turing, John von Neumann, Charles Babbage, Ada Lovelace. Then he stands perfectly still, this priestly young man, arms outspread in a cruciform posture.

"Around me shines the bits," he says, "and in me is the bytes. The data, the code, the communications. Forever, amen."

This young man, I learned, was a Swedish academic named Anders Sandberg. I was fascinated by the explicitness of Sandberg's curious ritual, its cultic acting out of transhumanism's religious subtext, but could not accurately gauge how seriously to take it—whether the performance was partly playful, partly parodic. I nevertheless found the scene strangely affecting, even haunting.

And then, shortly after I watched the documentary, I learned about a lecture Sandberg was due to deliver at Birkbeck College on the topic of cognitive enhancement. I made plans to go to London. It seemed as good a place as any to begin.

An Encounter

IT OCCURRED TO me, as I established myself in the back row of a packed lecture hall in Birkbeck and took brisk stock of the assembled crowd, that the future, such as it was, looked a lot like the past. Dr. Anders Sandberg's lecture had been organized by a group called London Futurists, a kind of transhumanist salon that had been meeting regularly since 2009 to discuss topics of interest to aspiring posthumans: radical extension of life spans, mind uploading, increased mental capacity through pharmacological and technological means, artificial intelligence, the enhancement of the human body through prostheses and genetic modification. We had gathered here to contemplate a profound societal shift, a coming transfiguration of the human condition, and yet there was no ignoring the fact that we were an overwhelmingly male group. Aside from the fact that almost all these faces were lit by the pallid luminescence of smartphone screens, this could be happening at almost any point in the last two centuries: a group, composed of mainly men, arranged in tiered seating in a room in Bloomsbury, there to listen to another man talk about the future.

A middle-aged gentleman with vigorous red eyebrows approached the lectern and took command of the room. This was David Wood, the chair of the London Futurists—a prominent transhumanist and tech entrepreneur. Wood had been a founder of Symbian, the first mass-market smartphone operating system, and his company Psion had been an early pioneer in the handheld computer market. He talked, in a meticulous Scottish accent, of how the next ten years would see more "fundamental and profound changes to the human experience than in any preceding ten-year period in history." He talked about the technological modification of brains, the refinement and enhancement of cognition itself.

"Can we get rid," he asked, "of some of the biases and mistakes in reasoning that we've all inherited from our biology? Instincts that served us well when we roamed the African savanna, but which are not now very much in our favor?"

The question seemed to encapsulate the transhumanist worldview, its conception of our minds and bodies as obsolete technologies, outmoded formats in need of complete overhaul.

He introduced Anders, who was these days a futurist of note, and a research fellow at Oxford's Future of Humanity Institute—an organization, founded in 2005 with an endowment from the tech entrepreneur James Martin, where philosophers and other academics were charged with conjuring and thinking through various scenarios for the future of the human species. Anders was recognizable still as the priestly young man whose strange and solitary observance I'd watched in the YouTube video, but he was in his early forties now, a fleshier and more substantial figure, adhering more or less to the disheveled house style of professional scholarship—the rumpled suit, the air of abstracted conviviality.

He spoke for the better part of two hours on the topic of intelligence, on how it might be increased at the level of the individual, and at the level of the species. He spoke of methods of cognitive augmentation, both existing and imminent—of education, smart drugs, genetic selection, brain implant technologies. He spoke of how as humans age, they lose their capacity to assimilate and retain information; life extension technologies, he allowed, would go some way toward addressing this situation, but we would also need to improve how our brains functioned through the course of our lives. He spoke of the social and economic costs of suboptimal mental performance, of how misplaced house keys alone—the time and energy invested in trying to find them—ran the U.K.'s GDP a deficit of £250 million every year.

"There are a lot of these little losses in society all the time," he said, "because of stupid mistakes, forgetfulness, and so on."

This struck me as an extreme manifestation of positivism. Anders spoke of intelligence as essentially a problem-solving tool, a function of productivity and yield—as something closer to the measurable processing power of a computer than any irreducibly human quality. In a general sense, I was fundamentally opposed to this conception of the mind. And yet in a personal sense I could not help but reflect on the fact that I myself had, through my own absentmindedness, only that morning squandered about £150, having somehow managed to book a hotel room in London for the night before I'd arrived, and having subsequently had to fork out for another one. I had always been somewhat scattered and forgetful, but since becoming a father—and resulting, at least in part, from such early parental phenomena as interrupted sleep, general distraction, and too much time spent watching episodes of *Thomas and Friends* on YouTube—my processing power, my memory capacity, had begun noticeably to decrease. And so as much as I was temperamentally resistant to the profoundly instrumentalist view of human intelligence Anders was advancing in his lecture, I couldn't help but feel that I could probably stand a little enhancement myself.

The thrust of his lecture was that biomedical cognitive enhancements would facilitate improved acquisition and retention of mental ability, of what he referred to as "human capital," allowing for better reasoning and functioning in the world. He addressed the questions of social justice that arose from this—questions of what he called "the fair distribution of brains"—given that those in a position to afford enhanced brains were likely to be those people already occupying an elite position within society. His suggestion, though, was that less intelligent people would wind up benefiting more from enhancement technologies than those who were already very intelligent, and that the overall effects of increased general intelligence would inevitably benefit society as a whole—a kind of trickle-down economics of intelligence.

All of this—the setup, the situation—was utterly familiar to me, and yet utterly strange. I had lately abandoned the sinking ship of an academic career for the hardly less precarious vessel of freelance writing. I had used up several years of my unextended life span getting a PhD in English literature, only to confirm my suspicion that a PhD in English literature was never going to lead me to the promised land of actual employment. I had spent much of my twenties and thirties trying to pay attention to people standing at lecterns and saying things. And yet the sorts of things that Anders Sandberg was saying were very different to the sorts of things I was used to hearing from people standing at lecterns. I was, yes, sitting in the back of a lecture hall and trying to focus on the matter at hand, an activity in which I was deeply and intricately experienced. But in no sense was I among my people. In no sense was this my world.

After the lecture, a contingent of assorted futurists migrated to an oak-paneled pub in Bloomsbury for some early afternoon drinking. By the time I sat down at the table with my pint of bitter, word had spread around the group that I was writing a book on transhumanism and related matters.

"You're writing a book!" said Anders, apparently delighted by the idea. He pointed to a hardback volume that sat in front of me on the table, a cultural history of severed heads I had been carrying around with me since I'd acquired it earlier that day. "Is that the book you're writing?"

"What, this?" I said, unsure as to whether I was missing some intricate transhumanist joke about cryonic head storage, or possibly time travel.

"No, that one's already been written," I said, unnecessarily. "I'm writing a book about transhumanism and related topics."

"Ah, excellent!" said Anders.

I wasn't sure what to say. I almost told him that the book I was planning to write might not be the sort of book that he, or transhumanists in general, would believe to be excellent. I felt suddenly conscious of myself as an interloper among these rationalists and futurists, an odd and perhaps even slightly pitiful figure, with my antediluvian notebook and pen, an emissary of letters in the world of zeros and ones.

I noted that Anders wore a pendant around his neck, a thing with a large medallion not unlike those devotional medals worn by especially pious Catholics. I was about to ask him about it when his attention was seized by an attractive Frenchwoman who wanted to talk about brain uploading.

An aristocratic young man who had been sitting to my left now turned toward me and asked about this book I was writing. He was elegantly attired, his hair punctiliously crafted. His name was Alberto Rizzoli, he told me, and he was from Italy. (At one point, in reference to my book, he mentioned that his family used to be in the publishing business. Only later that evening, as I glanced through my notes, did it occur to me that Alberto was surely a scion of the Rizzoli media dynasty, which would make him the grandson of Angelo Rizzoli, who had produced Fellini's *La Dolce Vita* and *8½*.) He was studying at the Cass Business School in London, but was also working on a beta-stage tech start-up, which provided 3D printing materials for primary classrooms. He was twenty-one years old, and had considered himself a transhumanist since his teens.

"I certainly can't imagine myself at thirty," he said, "without some kind of enhancement."

I myself was thirty-five, like Dante at the time of his vision—midway upon the journey of my life. And I was, for better or worse, unenhanced. As disturbed as I was by the idea of the cognitive augmentations Anders had spoken of in his lecture, I was nonetheless intrigued by the thought of what such technologies might do for me. Such technologies might, for instance, have freed me from the burden of having to take notes while talking to transhumanists, allowing me instead to record everything through some internal nanochip for purposes of later perfect recall, as well as, say, furnishing me with the extra-contextual information–in, as it were, real time–that the grandfather of the young Italian man I was speaking with had produced a bunch of Fellini films.

A silver-haired man in a sport coat and expensive-looking shirt sat down across from Alberto and me. He had positioned himself snugly beside Anders, and was waiting for a gap in his conversation with the Frenchwoman. In the meantime, he had helped himself to a couple of pistachios from Anders's snack bowl, one of which he had fumbled on the way to his mouth and dropped down the neck of his shirt, open to the ideally entrepreneurial three-to-four buttons. I watched him as he hooked a finger through a gap between two lower buttons and poked around momentarily before capturing the truant pistachio and popping it discreetly into his mouth. Our eyes met as he did so, and we smiled blandly in each other's direction. He handed me a card, from which I learned that he was in the professional futurism business. (I considered making a lighthearted joke about how a business card, attractive as this particular one was, seemed an oddly old-fashioned method for a professional futurist to be announcing his status as such, but I thought better of it, and crammed the card into the section of my wallet that served as the somewhat overcrowded final resting place of such printed disjecta.)

He had started out in artificial intelligence research, he said, but now made his living as a keynote speaker at business conferences, informing corporations and business leaders of trends and technologies that were going to disrupt their particular sectors. He spoke as though he were doing a brisk and slightly distracted run-through of a TED talk; his physical gestures were both emphatic and relaxed, suggesting a resolute optimism toward a horizon of vast and terrible disruptions. He spoke to me of those changes and opportunities that were at hand, of a near future in which AI would revolutionize the financial sector, and in which a great many lawyers and accountants would become literally redundant, their expensive labor made superfluous by ever smarter computers; he spoke to me of a future in which the law itself would be inscribed in the mechanisms through which we act and live, in which cars would automatically fine their drivers for breaking speed limits: a future in which there would in fact be no need for such things as drivers, or car manufacturers, given that vehicles would soon be sailing calmly out of showrooms like ghost-ships, still warm from the 3D printer from which they had lately emerged, according to the precise specifications of the consumer for whose home or workplace they were now setting course.

I told him that the one reassuring aspect of my job as a writer was that I was unlikely to get replaced by a machine anytime soon. I might not make a lot of money, I admitted, but I was at least in no immediate danger of being ejected outright from the marketplace by a gadget that did exactly what I did, but more cheaply and efficiently.

The man tilted his head from one side to the other, pursing his lips, as though considering whether to permit me this limited self-consolation.

"Sure," he conceded. "I mean certain kinds of journalism will probably not be replaced by AI. Opinion writing, in particular. People will probably always want to read opinions generated by actual humans."

Although hot takes were under no immediate threat, certain plays and films and works of prose fiction had, he said, already been written to order by computer programs. It was true that these plays and films and works of prose fiction were not very good, or so he had heard, but it was also true that computers tended to improve very quickly at things they initially did not do well. His point, I supposed, was that I and people like me were just as expendable as everyone else, just as fucked by the future. I considered asking him whether he thought computers might eventually replace even keynote speakers, whether the thought leaders of the next decade might fit in the palms of our hands, but realized that whatever answer he provided to this question would be cause for smug vindication on his part anyway, and so I resolved instead to include a description in my book of his retrieving a dropped pistachio from inside his expensive shirt—an act of petty and futile vengeance, and the kind of absurd irrelevance that would certainly be beneath the dignity and professional discipline of an automated writing AI.

Anders and the attractive Frenchwoman to my right were engaged in what seemed to me an impenetrably technical discussion about the progress of research into mind uploading. The conversation had turned to Ray Kurzweil, the inventor and entrepreneur and director of engineering at Google who had popularized the idea of the Technological Singularity, an eschatological prophecy about how the advent of AI will usher in a new human dispensation, a merger of people and machines, and a final eradication of death. Anders was saying that Kurzweil's view of brain emulation, among other things, was too crude, that it totally ignored what he called the "subcortical mess of motivations."

"*Emotions!*" said the Frenchwoman, emotionally. "He doesn't need emotions! That is why!"

"That might be true," said Alberto.

"He wants to become a machine!" she said. "That is what he really wants to be!"

"Well," said Anders, poking thoughtfully among the bowl of empty shells, searching in vain for an uneaten pistachio. "I also want to become a machine. But I want to be an *emotional* machine."

When I finally spoke at length with Anders, he expanded on this desire of his to become a machine, this literal aspiration toward a condition of hardware. As one of the foremost thinkers within the transhumanist movement, he was known as much as anything for his advocation and theorizing of the idea of mind uploading, of what was known among the initiates as "whole brain emulation."

It wasn't, he insisted, that he wanted this right away; even if such a thing might be possible in the near future–and he stressed that we were nowhere close–it wouldn't be desirable for humans to start getting uploaded into machines all of a sudden anyway. He spoke of the potential dangers of the sort of sudden convergence that techno-millenarians like Kurzweil refer to as the Singularity.

"What would be a nice scenario," he said, "is that we first get smart drugs and wearable technologies. And then life extension technologies. And then, finally, we get uploaded, and colonize space and so on." If we managed not to extinguish ourselves, or to be extinguished, what we now think of as humanity would be the nucleus, he believed, of some greatly more vast and brilliant phenomenon that would spread across the universe and "convert a lot of matter and energy into organized form, into life in a generalized sense."

He had held this view, he said, since childhood, since consuming wholesale the contents of the Stockholm municipal library's sci-fi section. In high school he read scientific textbooks for pure diversion, and kept a scrapbook of equations he found especially stimulating; he was excited, he said, by the movement of the logic, the lockstep progression of the thought–by the abstract symbols themselves more than the actual things they signified.

One especially rich source of such equations was a book called *The Anthropic Cosmological Principle* by John D. Barrow and Frank J. Tipler. At first, Anders read the book primarily for these tantalizing calculations–"weird formulas," as he put it, "about things like electrons orbiting hydrogen atoms in higher dimensions"–but like a kid with a copy of *Playboy* who eventually turns his attention to a Nabokov story, he began to take an interest in the text that surrounded them. The view of the universe advanced by Barrow and Tipler was as an essentially deterministic mechanism, in which "intelligent information processing must come into existence," and increase exponentially over time. This teleological premise led Tipler, in his later work, to the idea of the Omega Point, a projection whereby intelligent life takes over all matter in the universe, leading to a cosmological singularity, which he claims will allow future societies to resurrect the dead.

"The idea was a revelation to me," Anders told me. "This theory that life will eventually control all matter, all energy, and calculate an infinite amount of information—that was kind of awesome for an information-obsessed teenager. This was something I realized we needed to work on."

And this realization was, he said, the moment when he became a transhumanist. If the goal was to maximize the amount of life in the universe, and thereby the amount of information that would be processed, it was clear to him, he said, that humans needed to expand into the outer reaches of space, and to live for an extremely long time. And for those things to become a reality, it was clear that we were going to need AI, and robots, and space colonies, and certain other things he had read about in those sci-fi books in his local library.

"What is the value of a star?" he asked, and did not pause for an answer. "A star in itself is kind of interesting, if you have just one of them. But if you have trillions of them? Well, they are actually fairly alike. There is very little structural complexity there. But *life*," he said, "and in particular the life of individuals—that is highly contingent. You and I have a life story. If we reran the story of the universe, you and I would end up as different human beings. Our uniqueness is a thing we accumulate. That is why the loss of a person is something very bad."

Anders's vision of *getting uploaded*, of the conversion of human minds into software, was central to this ideal of transcending human limitations, of becoming a pure intelligence that would spread throughout the universe. In many ways, he seemed very different to the person I'd seen in that documentary, the slightly chilling figure making priestly gestures of technological benediction; he seemed not just older, but less machinelike, more fascinatingly human in his desire to be a machine.

But the visions of the future he was outlining were overwhelmingly strange and unsettling to me—more alien, and alienating, than any of the actual religious ideas I did not subscribe to, precisely because the technological means to their realization were at least theoretically within reach. Some essential element within me reacted with visceral distaste, even horror, to the prospect of becoming a machine. It seemed to me that to speak of colonizing the universe—of putting the universe to work on our projects—was to impose upon the meaningless void the deeper meaninglessness of our human insistence on meaning. I could imagine no greater absurdity, that is, than the insistence that everything be made to mean something.

The pendant that Anders wore around his neck—the silver medallion that looked very much like a Catholic religious medal, and which added to the overall clerical impression he gave off—was in fact etched with instructions for cryonic suspension of his earthly body in the event of his death. This, I understood, was a wish he held in common with a great many transhumanists: that their bodies would be preserved upon death in liquid nitrogen, against the day when some future technology might allow their thawing and reanimation, or when the 3.3 lbs of neural wetware inside their skulls might be removed, scanned for their repository of information, converted into code, and uploaded into some new type of mechanical body, not subject to decrepitude or death or other human defects.

The place where Anders's earthly body was to be sent, according to the instructions etched on the medallion, was a facility in Scottsdale, Arizona, called Alcor Life Extension Foundation. And it turned out that the man who ran this cryonics facility was Max More, the same Max More who had written the "Letter to Mother Nature." Alcor was where transhumanists went when they died, so that their deaths might not be irrevocable–where abstract concepts of immortality were brought into the physical realm. And I myself wanted to go there to be among those suspended immortals, or at any rate their frozen corpses.

A Visitation

IF YOU FLY to Phoenix, and then drive north for half an hour or so across a landscape reclaimed from the radiant emptiness of the Sonoran Desert, you will reach a squat gray block of a building, constructed for the purpose of preparing and storing bodies very much like your own for their eventual return to life. If you press the buzzer, and if someone lets you in, you will enter a vestibule decorated in a manner you might find suggestive of a mid-1990s straight-to-video sci-fi film—gleaming metallic-feature walls and chrome-effect furniture, all suffused in a gentle blue luminescence—and you will be invited to sit on a long and angular couch, there to await your guide to the afterlife.

In front of you, on a glass coffee table, you will find a slim volume you might care to flip through as you wait: a copy of an illustrated children's book called *Death Is Wrong*, the cover of which depicts a small boy scowlingly directing an index finger toward the eponymous leveler, with his hooded robe, his scythe, his terminally grinning skull. You will note, as you wait, the silence of this place, the absence of buzzing phones and humming printers and chattering staff, of all the ambient discourse you would expect in a typical place of business. It is possible that for long stretches of time, the only sounds you will hear will be the low whine of light aircraft taking off and landing at Scottsdale Municipal Airport, beside which this building, the headquarters of Alcor Life Extension Foundation, is conveniently located for the efficient delivery of the freshly dead.

Alcor is the largest of the world's four cryopreservation facilities—three of which are in the United States, and one in Russia. (It is no coincidence that these are the two countries whose narratives of national destiny were most firmly tethered, for much of recent history, to the exploration of space, and whose diametrically opposed ideologies were so driven by the notion of scientific progress.) Hundreds of people now living have made arrangements for their bodies to be brought here, as soon as possible after the pronouncement of clinical death, in order that certain procedures—including, as often as not, the removal of the head from the body—may be carried out, enabling their cryonic suspension until science figures out a way to bring them back to life.

Of Alcor's client base, there is a small contingent, currently numbering 117, who are no longer among the quick; they are known as "patients"—not as bodies, or as corpses, or as severed heads—because they are considered to be *suspended,* rather than deceased: detained in some liminal stasis between this world and whatever follows it, or does not. It was these suspended souls I myself had come to the desert suburbs to be among.

And I had come, too, to meet with Max More, who, as well as being a self-proclaimed founder of the movement, was also the president and CEO of Alcor. I wanted to know how a man who had ostensibly dedicated his life to the overcoming of human frailties, to a resolute transgression of the principle of entropy, had come to spend his days surrounded by corpses in an office park, between a tile showroom and a place called Big D's Floor Covering Supplies.

But first I wanted to know what actually went on in this place, what was done to the bodies of Alcor's clients in order to forestall their decay, their eventual loss to time. Max, a hulking presence in a tight-fitted black T-shirt, walked me down a narrow hallway toward the room where the patients were processed, and informed me that much of this had to do with which of the two main payment options you went for. For $200,000, Alcor would keep your entire body suspended until such time as it might once more be of some use to you; for $80,000, you could become what was known as a "neuro-patient," whereby your head alone—detached, petrified, chambered in steel—would be cryopreserved, with a view to the later uploading of your brain, or your mind, into some kind of artificial body.

In the past, the clients' estates, or their families themselves, would have covered these costs, paying out regular installments after their loved ones' deaths, but this was soon found to be impractical, as there were cases where the families could not make the payments, or simply saw no compelling reason to do so, and stopped—at which point you were left with basically an orphaned corpse, with no one to cover the cost of its suspension and eventual awakening. And so these days, Alcor's clients typically paid their bills through their life insurance plans, having kept up annual membership dues through the course of their natural lives.

I was given to understand that Max himself was a neuro man, despite the evidently significant investment he had made over the years in bulking and sculpting his limbs and torso. (Max was a kind of physical embodiment of his own ideals: gym-hewn, vigorous, controlled in his movements. His red hair had thinned and retreated toward the high ground of his crown, and this further dramatized the effect of the domed forehead, the strenuous vectors of the brows, the pale and illegible eyes.) His reasoning here, he said, was that he planned to stick around for another forty years or so, by which point, no matter how much of the intervening time he spent lifting weights, his body would likely not be worth retaining. Because part of what you're banking on with the neuro deal is the scientists of the future finding a way of providing reconstituted brains with new bodies, whatever form those might take.

Although it is the patient's brain in which Alcor is primarily interested, it is not the organization's practice to remove that brain from its casing of bone, its personal integument of muscle and skin—because the skull acts as a useful ready-made casing for the brain, providing extra protection during the period of cryopreservation, and because, technically speaking, it is kind of a hassle to remove the thing entirely, what with all the tissues and ligaments and so on connecting it to the interior of the cranium.

Max's manner was smoothly clinical: a GP talking a patient through a procedure. Making calm, steeple-fingered inventory of the benefits, the potential side effects. *Ask your doctor if immortality is right for you.*

The scientific basis for all of this was thin—was, in fact, essentially nonexistent. The promise of cryonics was purely theoretical: that science might one day advance to the point where it would be possible to thaw these bodies, these heads, and somehow reanimate them, or digitally duplicate the minds they contained. And this was all so speculative, so remote from anything that might be achievable, that the scientific community as a whole regarded it as barely worth the trouble of refutation. Those who did comment on it tended to do so with outright contempt. Writing in the *MIT Technology Review*, for instance, the McGill University neurobiologist Michael Hendricks insisted that "reanimation or simulation is an abjectly false hope that is beyond the promise of technology," and that "those who profit from this hope deserve our anger and contempt."

Near the entrance to the intake room, lying in an open-casket-like container made of lightweight canvas and filled with plastic imitation ice cubes, was the smooth-bodied likeness of a youngish white male—a mannequin laid low in its prime, its expressionless face largely covered by a respirator mask. This serene figure was intended as a stand-in for the potential client, the still living subject who had come to have explained to him those things that would be done to his body in the minutes and hours after its clinical death, should he choose to become a full member.

The preferred situation, Max told me, was that in which the client's clinical death occurred in a relatively predictable fashion, so that

Alcor's standby personnel could be present at the scene in time to begin the process of cooling the body, before its final journey, by air or road, to Phoenix.

The success of the procedure depends, to a large extent, on the predictability of the death. So cancer, on aggregate, is good: if you want a strong chance of an extended life span, terminal cancer is an excellent place to start from. A heart attack is less good, because it's extremely difficult to predict when it's going to get you. An aneurysm or a stroke is worse again, because if it's strong enough to kill you, it's likely to leave you with brain damage, and that's going to be tricky to deal with down the line—although not, of course, impossible, because we are talking here about future science. Accidents and other disasters are really at the bottom of the scale. There wasn't much that could be done, for instance, with the body of the Alcor client who died in the World Trade Center on September 11, 2001. More recently, another member died in a plane crash in Alaska.

"That was not ideal," Max informed me, his face a death-mask of irony.

If you're a whole body patient, you, or your whole body, will be placed on a tilted operating table, surrounded on four sides by sheets of Perspex. Small holes will then be drilled into your skull, so that your cryonics team can judge the condition of your brain, observe its state of swelling or contraction. Then they will open up your chest to gain access to your heart, connecting your major veins and arteries to a bank of perfusion machinery, so that your blood and bodily fluids can be flushed out and replaced as quickly as possible by a cryoprotectant agent—"Kind of like a medical-grade antifreeze, if you like"—which will guard against the formation of ice crystals. If you want to be preserved in any kind of reasonable shape long enough for future science to restore you to life, you don't want ice crystals forming in your cells. Ice crystals are high on the list of things that will seriously fuck up your post-resurrection quality-of-life prospects.

"So what you want to do," said Max, "is vitrify, rather than freeze. Vitrification forms a kind of resinous block that just holds everything in place. No sharp angles and edges."

If you're a neuro-patient, there is the matter of your decapitation to be attended to. This procedure is performed on the operating table. In the technical vernacular of cryonics, your detached head is referred to as the *cephalon*. (This, I later learned, was a primarily zoological term, referring to the head section of segmented arthropods such as marine-dwelling trilobites. Why this term was deemed preferable to "head" I could not begin to say, apart from that it deflected attention away from the fact that what we were talking about was severed heads—a deflection that was, I felt, not entirely successful.) And this cephalon, once removed from the body, is taken to a Perspex container known as "the cephalon box," inside of which it is held in an inverted position by a circular arrangement of clamps, until such procedures have been carried out as will enable its cryonic suspension.

At no point during this tour did Max betray any acknowledgment of the strangeness of the things he was telling me; the morbid ritual of B-movie dismemberment was outlined as though it were a straight-forward matter of medical expediency—which, in the hopeful thana-tology of cryonics, is exactly what it is.

The 117 patients currently suspended at Alcor were located in what was known as the "patient care bay." This was a large and high-ceilinged warehouse, filled with eight-foot-tall stainless steel cylinders, each of which was branded with Alcor's blue and white logo. This logo was a stylized letter A. Its predecessor was a more richly emblematic image of a white human figure with raised arms, contained within the larger blue figure of a phoenix, wings aloft. (While we're on the topic, let's allow ourselves to dwell momentarily on the novelistic strangeness of this future-resurrection venture having wound up headquartered on the outskirts of a city named after this mythical desert bird, with its cyclical existence of immolation and renewal. It's a detail at which you, reader, would probably wrinkle your delicate and finely attuned nose if you came across it in a work of fiction. And you'd be right to, of course: it's obviously a bit much.)

The cylinders are referred to as *dewars;* they are, essentially, gigantic thermos flasks filled with liquid nitrogen, each of which contains sufficient room for four whole-body patients, in a circular arrangement of compartments around a central column in which several cephalons can be stacked. The individual patients are stored in sleeping bags inside aluminum pods. Cephalon-only dewars, Max told me, could store as many as forty-five severed heads, each of which is contained inside a small metal cylinder, resembling the sort of stainless steel wastepaper basket you'd pick up in the bathroom section of IKEA. (Storage costs are the main reason it's cheaper to sign up as a neuro-patient than a whole body patient.)

As we walked among the shadows of the towering dewars, I tried to imagine the suspended bodies and heads within, this rigid delegation of the dead awaiting its chance to partake of a world to come. I knew that the body of a man named Dick Clair, the creator of the 1970s sitcom *The Facts of Life,* had been contained in one of these dewars since his death from AIDS in 1988. As was the head of the baseball legend Ted Williams. And I understood, too, that we were in the presence of the remains of the writer known as FM-2030—the Iranian futurist who legally changed his name from Fereidoun M. Esfandiary to reflect his conviction that the problem of human mortality would be solved by the year 2030—though I could not say precisely which dewar contained them, because for security reasons members of the public were not to be informed of the specific resting places of individual cryonauts. Max had mentioned to me that his wife, Natasha, had been involved with FM-2030 when they first met, and so I became briefly preoccupied, there in the care bay, with the gothic richness, as an idea, of this man charged with the maintenance of the corpse of his wife's former lover, a techno-utopian who believed in his own exemption from death.

But to reiterate: for Max, as for anyone who signs up to cryonics and its taxonomy, these are by no means corpses.

"Cryonics," as he put it, "is really just an extension of emergency medicine."

It would be easy, in the light of this seemingly bald denial of clinical orthodoxy, to portray cryonics as some kind of cult, or to view this place as a satirical diorama on the theme of modern scientism and its tragicomic excesses. But no one here is claiming that your return to life is assured if you simply sign up. Max himself admits that the whole setup is a Hail Mary pass into the end zone of the future. But the key sales pitch here is that it's got to be at least worth a shot, because although you may not be guaranteed resurrection if you sign up, you're seriously diminishing your chances if you don't. (You would not by any means be the first person to think here of Pascal's Wager.)

"Personally," said Max as we made our way through the patient care bay toward the exit, "I'm hoping to avoid having to be preserved. My ideal scenario is I stay healthy and take care of myself, and more funding goes into life extension research, and we actually achieve longevity escape velocity." He was referring here to the scenario, projected by the life extension impresario Aubrey de Grey, a scientific advisor at Alcor, whereby for every year that passes, the progress of longevity research is such that average human life expectancy increases by more than a year–a situation that would, in theory, lead to our effectively outrunning death.

"Of course, I could get hit by a truck," Max said. "Or someone could murder me. But the idea of sitting in one of those tanks, not being in control of my own destiny, doesn't actually appeal to me very much. It's just that it's obviously better than the alternative."

Laid flat on the floor by the entrance to the patient care bay was a dewar much smaller, and much older, than the others. It was open at one end, so that its narrow interior tube was visible. At the other end, a plaque announced that this was the very dewar in which one James H. Bedford, PhD, was originally contained before he was relocated here from Southern California, and moved to a more modern container in 1991. Bedford, a University of California psychology professor, was (or is) the first human ever to be cryopreserved. His preservation was performed in 1966 by a chemist, a physician, and a television repairman from L.A. named Robert Nelson, who was involved in his capacity as president of the Cryonics Society of California.

Max mentioned casually that because Bedford was born in 1893, this technically made him the world's oldest living person. I suggested that it was a bit of a stretch to call him living; Max suggested that it was not.

The idea, he reminded me, was that these patients were resuscitated shortly after the point of legal death, and cryopreserved when their bodies were still undecayed. It was a central premise of cryonics that *real* death, actual death, occurred not when the heart stopped beating, but several minutes later, when the body's cell and chemical structures began to disintegrate to the point where no technology could restore them to their original state. And so these cryopreserved corpses were not by conventional standards deceased—were not, that is to say, corpses at all—but rather human beings preserved between conditions of life and death, abiding in some state outside of time itself.

And standing in the cool of the care bay, surrounded by the unseen bodies and severed heads of techno-utopians, I thought of the Catholic concept of limbo, a place that was neither heaven nor hell, but a state of suspension, a holding pattern for the souls of the righteous who had died before they could be properly redeemed by the coming of Christ, and must wait in ontological détente for that day of salvation.

Here in the Sonoran Desert, I thought, protected by stainless steel receptacles and Kevlar walls and bulletproof glass, these patient souls were being held in a state of hopeful deferral, until the future came to deliver them from their own deaths. These men and women, these bodies and heads, would almost certainly never be returned to life, yet there was something inscrutably sacred in their suspension, in their waiting. This warehouse—mausoleum though it was of modern delusions—was at the same time a site of something ancient and primal. I was standing, I felt, on consecrated ground, in a place that was neither here nor there.

But no, I thought, that wasn't quite true, because I was very much in a specific place called America. I was here, in the old open ground of the colonial frontier, the theater of westward expansion in which the American drama of boundless national potential and individual fulfillment—the vast blood-and-gold fantasia of Manifest Destiny—was first enacted. The scene in which I was standing, with its immense silver canisters and its intricate display of gadgetry, began to seem a crazed pageantry of technological ingenuity and control, a sci-fi film set that might suddenly be dismantled and carried away, leaving nothing but the desert of the old American West, which had always been a landscape of death.

I imagined a delegation of explorers from some civilization of the distant future excavating these dewars from the depths of the desert, and looking with detached fascination at the half-preserved remains within, the bodies, the cephalons, puzzling over who these people were and what it was that they believed. And I wondered how I might answer their questions, if I could somehow do so. Would I say that they believed in science? That they believed in the future? That they believed in never growing old? That they believed in their life insurance policies? That they believed in the mysterious power of applied money? That they believed in themselves? That they were, in one way or another, Americans?

Alcor's mission is presented as a humanitarian one: like any business, they want to expand their customer base, but this objective also happens to be theoretically aligned with the overall aim of defeating death. A rising tide lifts all boats, is the idea. There's a long article on the company's website about what it would actually involve, at a technical level, to ensure through cryonic suspension the future resurrection of every person now living. The article, called "How to Cryopreserve

Everyone," is by the computer scientist Ralph Merkle, the inventor of public key cryptography. Merkle describes Alcor's animating principle as a "vision of the future in which everyone alive today can enjoy good health and long life in a world of material abundance for all." What we know for certain, the author asserts, is that "advances in technology will eventually make this future a reality."

But it's not like anyone is saying there aren't some kinks to be worked out here. The financial costs of suspending all these patients would be one issue, certainly, as would the brute geometry of storage: where do you put them all, the bodies of those named in the book of life? Actually, we'd be talking not so much about bodies as heads, because the logistics of whole body preservation for every living person would be truly nightmarish, as opposed to merely problematic. Merkle's article proposes, as a potential solution to this difficulty, the idea of Really Big Dewars (RBDs).

The annual global mortality rate, Merkle writes, is somewhere in the region of 55 million. Now let's say we build a gigantic spherical dewar thirty meters in radius. Given the dimensions of the average human head, this thing would comfortably accommodate 5.5 million cephalons; and so if you built ten of these RBDs per annum, you'd have the means to store the head of every person who died in the entire world, going forward, until such time as their deaths could be remedied.

There would, naturally, be significant costs attached to all of this. Each of these RBDs would have a volume of about 113 million liters, which means your outlay for liquid nitrogen—which tends to come in at 10 cents or so per liter—would be about $11 million per RBD. There'd also be some added expenses associated with boil-off rate and insulation and general dewar maintenance, but ultimately the cost of cryopreserving the entire population of earth would come in at a surprisingly competitive amortized capital sum of $24–$32 per literal head. (For whole body patients, we'd be looking at putting a zero at the end of those figures.)

The point is that cryonics, as both a business and a tactic for evading the fate that awaits us all, is an at least theoretically scalable model.

Alcor was a place built to house the corpses of optimists; the silence there was thick with ironies. And the irony with which I found myself most immediately preoccupied was the situation of Max himself, or the picture of that situation I could not help but cultivate in my mind.

This was a man who had dedicated his life to the idea of transcending the limitations of our natural condition, to a vast expansion of the range of human experience and potential. This was a man who, before he left Britain for the U.S. in his twenties, started the Extropian movement, named in defiance of an entropic principle whereby all that exists tends toward disintegration and disorder and decline, in a universe in which the center cannot hold. This was a man who had dedicated himself to what he called "perpetually overcoming constraints on our progress and possibilities as individuals, as organizations, and as a species." This was a man who changed his name, in a youthful gesture of radical self-invention, from Max O'Connor to Max More because, as he once put it in an interview with *Wired* magazine, "it seemed to really encapsulate the essence of what my goal is: always to improve, never to be static. I was going to get better at everything, become smarter, fitter, and healthier. It would be a constant reminder to keep moving forward."* This was a man who had undertaken, in an explicit and sustained fashion, the Nietzschean task of self-overcoming.

* For what it's worth, he gave a slightly different reason for the name change at the time, in an announcement about it in the Summer 1990 issue of *Extropy* magazine, the house publication of the Extropian movement: "I am no longer 'Max O'Connor.' I've changed my name to 'Max More' in order to remove the cultural links to Ireland (which connotes backwardness rather than future-orientation) and to reflect the extropian desire for MORE LIFE, MORE INTELLIGENCE, MORE FREEDOM."

And yet here he was, this man, spending his days in a small office in a kind of industrial estate in a suburb of Phoenix, surrounded by the dead. He was a cultivator of hope, it was true, but he was also a processor of bodies, a custodian of corpses: an executive-level necrocrat.

In his introduction to a recently published anthology called *The Transhumanist Reader,* which he had edited with his wife, Natasha, Max wrote the following: "Becoming posthuman means exceeding the limitations that define the less desirable aspects of the 'human condition.' Posthuman beings would no longer suffer from disease, aging, and inevitable death."

Max's conviction that the technologies of the future would release us from our human deficiencies grew out of what seemed a kind of congenital optimism. (His mother, he claimed, named him Maximilian, meaning "Greatest," because he was the heaviest baby in the hospital ward where he was born.) It was, he felt, almost as though he was born with some kind of transhumanist gene. As far back as he could remember, it was always there in him—this hunger for transcendence, this yearning to overcome.

Growing up in the port town of Bristol, in the southwest of England, he was fascinated with space, with the idea of colonizing other worlds. "When I was five," he told me, "I watched the Apollo moon landings. I was one of the few people who stuck with it. I watched every landing after that. I just loved the whole idea of getting off this planet." He was devoted to the children's television show *The Tomorrow People,* which ran through most of the 1970s on British television, and which centered around a group of teenagers whose extraordinary capabilities—telepathy, telekinesis, teleportation—revealed them as a sort of advance guard of future human evolution. The teens were assisted in their world-saving ventures by an artificial intelligence named TIM, who was housed in a disused London tube station. Max haunted the science fiction sections of Bristol's bookshops and libraries; he read a great many superhero comics, too, and these last exerted a formative pressure on his developing sense of the possibilities for a human future. (Stan Lee's *Iron Man* comics, with their fantastic vision of a technologically enhanced human body, were a particular influence.)

By the age of ten or eleven, his precocious interest in human enhancement led to his dabbling in the occult mysteries of Rosicrucianism. By thirteen, he had moved on to the Jewish mysticism of the Kabbalah. At the otherwise very conservative boarding school he attended, his Latin teacher gave classes in Transcendental Meditation, and he was one of two boys who signed up. But he soon found that he lacked the temperament for meditation, for it its rigors of stillness and patience.

By his mid-teens, he was, as he put it, developing stronger critical thinking skills, and moving away from the more esoteric fascinations of his early adolescence. He discovered libertarianism—something that has remained a central strand of his thinking ever since—through reading *The Illuminatus! Trilogy* by Robert Shea and Robert Anton Wilson (although the novels in fact represented libertarian and Randian ideas only in order to ridicule them). And it was through Wilson, too, that he first learned about cryonics. In a book called *Cosmic Trigger I: Final Secret of the Illuminati,* Wilson wrote of his decision to cryonically preserve the head of his daughter, Luna, who was beaten to death in a robbery in the San Francisco clothes store where she worked.

Through a group called the Libertarian Alliance, which he'd joined after reading the *Illuminatus* books, Max became friendly with people whose interests extended to space colonization and the enhancement of human intelligence. Cryonics was a popular topic within this new circle of acquaintances, and he began to establish himself as a leading vector of the idea. In 1986, while still a student in economics at Oxford, he spent six weeks in California, on a sort of fact-finding mission at Alcor's original headquarters in Riverside. When he returned to England, he helped set up the first cryonics society outside America.

In 1987, after finishing his degree at Oxford, he moved to L.A., where he started a PhD in philosophy at the University of Southern California. His dissertation explored the nature of death, and the continuity of the self over time. This work was clearly drawing on his interests in cryonics and life extension, but whenever he tried to raise these issues directly with his advisor, she would become visibly uncomfortable.

"I'd ask her whether she thought it wouldn't work," Max said.

We were sitting, now, at an oval boardroom table, across from a large bulletproof window overlooking the vista of the patient care bay.

He said, "I wanted to know did she have philosophical objections, did she think it wouldn't be *you* if you were brought back to life, or if you had your mind uploaded? And she'd say 'No,' and I'd say, 'Well, what's the problem then?' And she'd say, 'The whole thing is just so *ghastly*!'"

As he spoke, he leaned forward in his leather boardroom chair, and an old frustration fleetingly revealed itself in the tensing muscular planes of his face.

"Well, it's hard to know what to say to that," he said. "Ghastly as opposed to what, exactly? Putting your body in the ground and having it slowly digested by worms and bacteria?"

He shook his head, then spread his hands in a gesture of stoic forbearance. This whole reflexive disgust thing, he said, was a real problem. Leon Kass, he said, the former chairman of the President's Council on Bioethics, had written a book called *Beyond Therapy* that was essentially a lengthy argument against transhumanism.

"Kass came up with this idea of 'the Wisdom of Repugnance,'" Max said, "which is basically if something feels wrong to him, then it *is* wrong. People have these kinds of instinctive reactions, based in all these myths that teach us to fear going beyond our limits. You know: the Tower of Babel; Prometheus stealing fire from the gods and getting his liver eaten by an eagle. But people will always think something is terrible when it's in the future. Once it's here, they'll accept it."

Early in his time at USC, he met a young law student named Tom Bell, a fellow libertarian who shared Max's vaulting optimism about subjects like life extension, intelligence augmentation, and nanotechnology. They started putting together a magazine called *Extropy: The Journal of Transhumanist Thought,* and soon afterward set up a nonprofit they called the Extropy Institute. Although Max is the figure most closely associated with Extropianism, which is generally seen as an early version of the transhumanist movement, it was Bell, he says, who coined the term. In those days, he went by the name T. O. Morrow, but since the late 1990s he has reverted to the less hurtlingly dynamic Tom W. Bell.

Max maintains that a document he wrote in 1990 called "The Extropian Principles"—laying out the movement's ideals of "Boundless Expansion," "Self-Transformation," "Dynamic Optimism," "Intelligent Technology," and "Spontaneous Order"—constitutes the "first comprehensive and explicit statement of transhumanism." The Extropy Institute lasted until the mid-2000s, at which point it became more or less absorbed by the broader transhumanist movement, which is at least notionally contained under the official institutional umbrella of a group called Humanity Plus, an organization chaired by Max's wife, Natasha Vita-More.

Max and Natasha met at a dinner party in the early 1990s. The party was hosted by the 1960s acid guru Timothy Leary, who was by that late stage of his life a committed advocate of cryonics and life extension.* Despite Natasha being a decade and a half Max's senior,

* In the 1970s, while incarcerated for a range of drug offenses, Leary developed a set of futurist principles under the snappy rubric SMI²LE (Space Migration, Intelligence Increase, Life Extension). He was a long-standing member of Alcor, active to the point of hosting, on several occasions, the foundation's annual turkey roast at his home; but when the time eventually came to make the necessary arrangements, he went for the more show-stopping option of having his cremated ashes shot into space from a cannon. This is still a sore point within the cryonics community, and is seen as a significant tragedy—a stance that, when you think about it, is entirely consistent with the immortalist worldview. In a 1996 issue of *Extropy* magazine, Max and Natasha criticized Leary's decision as a sad capitulation to "deathist" ideology.

there was an immediate attraction, and an intellectual connection, though Natasha was still at that point in a relationship with FM-2030. Six months later, when that relationship had finally ended, she invited Max to appear as a guest on a TV chat show she hosted on L.A. local cable, and they quickly began dating.

I visited Natasha at the minimally chic house she and Max shared with a pleasant if somewhat overfamiliar Goldendoodle named Oscar, who was getting on in years but had lately become the beneficiary of a pet-specific cryonic preservation policy. Natasha was eating a hurried late breakfast of muesli and fruit when I arrived, having just returned home from teaching an early class on futurism at the University of Advancing Technology, a private college in Tempe.

She was sixty-five, a composed and austerely elegant figure whose manner was alternately warm and watchful, her stern good looks remarkably well preserved against the advance of time. She spoke of her marriage to Max as a union of complementary opposites: a synthesis of the analytical and the artistic, the scholar and the socialite. She made much of his Englishness, of his undergraduate degree from Oxford, and of the fact that he was fifteen years her junior.

"We're from different generations," she said, "and we come from very different worlds."

Natasha spent the 1970s and 1980s moving between the worlds of avant-garde art and independent film. She ran a performance art nightclub on Sunset Boulevard; she wrote for *The Hollywood Reporter;* she worked for a time for Francis Ford Coppola; and she was, she indicated, acquainted in those years with such luminaries as Werner Herzog and Bernardo Bertolucci.

She spoke of this period of her life in long, free-associating riffs, dense with allusion to all manner of people, all manner of philosophies. She spoke of backing up the brain, backing up the body. She spoke of the weakness of the flesh, and the power of technology. She had the manner of a mystic, some crystal-gazing quality that was both intensity and absence, as though she was already speaking from the distant future.

Her name, like Max's, was the heraldic device of her commitment, her promise to herself. Vita-More. More life.

It was, she told me, through a frightening encounter with her own bodily frailty in her early thirties that she began to think seriously about technology and mortality. In 1981, she suffered an ectopic pregnancy, and lost the child she was carrying. When she was taken to the hospital, having been found in a spreading pool of her own blood, she was minutes from death. When she talked now of her path to transhumanism, this was the time of her life she continually returned to, the moment when she realized, on a visceral level, that the human body was a feeble and treacherous mechanism, that we were each of us trapped, bleeding, marked for death.

"People ask how it's possible to think freely if you live somewhere like North Korea, where the government is strictly controlling everything," she said. "But our personhood is bounded by this secretive and unknown thing, this body. After my illness, I started seeing things differently. I became very interested in human enhancement, in how we might protect ourselves from this tyrannical onslaught of disease and mortality."

In an essay on mind uploading, Max wrote of his intention, if he lived long enough, to "exchange my physical body for a choice of bodies both physical and virtual." The question of what these future vessels of being might look like, or how they might function, was wide open, but one possible answer took the form of Natasha's Primo Posthuman project. This was a blueprint for what she called a "platform diverse body," a kind of reductio ad absurdum of the logic of wearable tech, whereby the human form itself was entirely replaced by a sleekly anthropomorphic device—a "more powerful, better suspended and more flexible... body offering extended performance and modern style"—which would be inhabited and controlled by an uploaded, substrate-independent mind.

This was her prototype for the unfleshed future, her vision of a form that would one day accommodate the uploaded content of human minds—including her own, and Max's. The content of those detached heads in Alcor's dewars, those human lives in cold storage, awaiting return. This was Natasha's suggestion of how they might live again, in this gleaming anthrobot, with its nanotech storage system, its instant data replay and feedback, its embedded high-throughput contradiction detectors.

And wasn't Natasha's vision of a wholly mechanized body, of an impenetrable shell of technology, also a dream self-portrait—a creative denial of her own frailty and mortality?

"If this body fails," she said, "we have to have another one. You could die at any moment, and that's unnecessary and unacceptable. As a transhumanist, I have no regard for death. I'm impatient with it, annoyed. We're a neurotic species—because of our mortality, because death is always breathing down our necks."

I could not disagree. It had always been unacceptable, this condition, it had always been the cause of our estrangement from ourselves. Speaking to Natasha reminded me of what I had always found so disturbing about transhumanism. There was the truth of its premise, that we were all of us trapped, bleeding, marked for death. And there was the strangeness of its promise, that technology could redeem us, release us from that state. These things both did and did not connect.

These proposals—cryonic suspension, mind-driven avatars—seemed to hover on some unreal threshold between technological hope and mortal terror. I could not imagine placing my faith in them. But then I could not place my faith in the world in which I spent my life, the so-called real world with its improbable technologies, its economies and systems based on mass delusion, giddy suspensions of disbelief, unimaginable innovations and savageries. None of it was remotely plausible, as far as I was concerned, and yet here we were.

Such were my thoughts, at any rate, as I sat at the departure gate at Phoenix airport, waiting to board a flight to San Francisco. I was still jet-lagged from my flight from Dublin, still feeling half unreal, half displaced. Wasn't technology itself, I thought, a strategy for disembodiment? Wasn't it all—social media, Internet, air travel, space race, telegraph, railway, the invention of the wheel—an ancient yearning to be out of ourselves, out of our bodies, our location in space and time?

These thoughts were the outcome of the conversations I had had with Max and Natasha, and of the hours I had spent among cryonically preserved bodies, but also of the fact that I was about to meet, in San Francisco, a man whose goal was the final displacement of nature itself. I was on my way to see a neuroscientist whose long-term project was the exact future for which the cephalons of Alcor remained in suspended hope: the uploading of human minds into machines.

Once Out of Nature

HERE'S WHAT HAPPENS. You are laid on an operating table, fully conscious, but rendered otherwise insensible, otherwise incapable of movement. A humanoid machine appears at your side, bowing to its task with ceremonial formality. With a brisk sequence of motions, the machine removes a large panel of bone from the rear of your cranium, before carefully laying its fingers, fine and delicate as spiders' legs, on the viscid surface of your brain. You may be experiencing some misgivings about the procedure at this point. Put them aside, if you can. You're in pretty deep with this thing; there's no backing out now.

With their high-resolution microscopic receptors, the machine fingers scan the chemical structure of your brain, transferring the data to a powerful computer on the other side of the operating table. They are sinking further into your cerebral matter now, these fingers, scanning deeper and deeper layers of neurons, building a three-dimensional map of their endlessly complex interrelations, all the while creating code to model this activity in the computer's hardware. As the work proceeds, another mechanical appendage–less delicate, less careful–removes the scanned material to a biological waste container for later disposal.

This is material you will no longer be needing.

At some point, you become aware that you are no longer present in your body. You observe–with sadness, or horror, or detached curiosity—the diminishing spasms of that body on the operating table, the last useless convulsions of a discontinued meat.

The animal life is over now. The machine life has begun.

This, more or less, is the scenario outlined by Hans Moravec, a professor of cognitive robotics at Carnegie Mellon, in his book *Mind Children: The Future of Robot and Human Intelligence*. It is Moravec's conviction that the future of the human species will involve a mass-scale desertion of our biological bodies, effected by procedures of this kind. It's a belief shared by many transhumanists. Ray Kurzweil, for one, is a prominent advocate of the idea of mind uploading. "An emulation of the human brain running on an electronic system," he writes in *The Singularity Is Near*, "would run much faster than our biological brains. Although human brains benefit from massive parallelism (on the order of one hundred trillion interneuronal connections, all potentially operating simultaneously), the rest time of the connections is extremely slow compared to contemporary electronics." The technologies required to perform such an emulation—sufficiently powerful and capacious computers, and sufficiently advanced brain-scanning techniques—will be available, he announces, by the early 2030s.

And this, obviously, is no small claim. We are talking about not just radically extended life spans, but also radically expanded cognitive abilities. We are talking about endless copies and iterations of the self. Having undergone a procedure like this, you would exist—to the extent that you could meaningfully be said to exist at all—as an entity of unbounded possibilities.

I knew that this notion of disembodied mind was central to transhumanism. I knew that this final act of secession from nature was, in fact, the highest ideal of the movement, the very future for which all those bodies and heads were being preserved in giant dewars at Alcor. But it was my understanding that the concept remained squarely in the realm of speculation, a matter purely for sci-fi novels, techno-futurist polemics, philosophical thought experiments.

And then I met a man named Randal Koene.

I was introduced to Randal at a Bay Area transhumanist conference. He wasn't speaking at the conference, but had come along out of personal interest. He was a cheerfully reserved man in his early forties, and he spoke in the punctilious staccato of a non-native-English speaker who had long mastered the language. We talked only briefly, and I must confess that I was not at that point entirely clear on what it was that he did. As we parted, he handed me his business card, and it was only much later that evening, when I'd returned to my rented place in the Mission neighborhood of San Francisco, that I removed it from my wallet and had a proper look at it. The card was illustrated with a picture of a laptop, on whose screen was displayed a stylized image of a brain. Underneath was printed what seemed to me an attractively mysterious message: "Carboncopies: Realistic Routes to Substrate Independent Minds. Randal A. Koene, founder."

I took out my laptop and went to the website of Carboncopies, which I learned was a "nonprofit organization with a goal of advancing the reverse engineering of neural tissue and complete brains, Whole Brain Emulation and development of neuroprostheses that reproduce functions of mind, creating what we call Substrate Independent Minds." This latter term, I read, was the "objective to be able to sustain person-specific functions of mind and experience in many different operational substrates besides the biological brain." And this, I further learned, was a process "analogous to that by which platform-independent code can be compiled and run on many different computing platforms."

It seemed that I had met, without realizing it, a person who was actively working toward the kind of brain uploading scenario that Anders and Max and Natasha had spoken about, and which Ray Kurzweil had outlined in *The Singularity Is Near*. And this was a person I needed to get to know.

Randal Koene was an affable and precisely eloquent man, and his conversation was unusually engaging for someone so forbiddingly intelligent, and who worked in so rarefied a field as computational neuroscience; and so, in his company, I often found myself momentarily forgetting about the nearly unthinkable implications of the work he was doing, the profound metaphysical weirdness of the things he was explaining to me. He'd be talking about some tangential topic–his happily cordial relationship with his ex-wife, say, or the cultural differences between European and American scientific communities–and I'd remember with a slow uncanny suffusion of unease that his work, were it to yield the kind of results he is aiming for, would amount to the most significant event since the evolution of *Homo sapiens*. The odds seemed pretty long from where I was standing, certainly, but then again, I reminded myself, the history of science was in many ways an almanac of highly unlikely victories.

One evening in early spring, Randal drove down to San Francisco from the North Bay, where he lived and worked in a rented ranch house surrounded by rabbits, to meet me for dinner in a small Argentinian restaurant on Columbus Avenue. (The menu, as it happened, featured a dish named "Half Rabbit," and although Randal was tempted, he felt that he could not, in good conscience, enjoy such a dish in the knowledge that he would have to return home and meet the gaze of the whole rabbits with whom he shared an estate. He went with chicken instead.) He was dressed entirely in black–black shirt, black cargo pants, black shoes–with the jazzy exception of a bright green Nehru jacket, leaf-patterned and mandarin-collared, all of which lent him the somewhat self-contradictory and misleading appearance of a survivalist mystic.

The faint trace of an accent turned out to be Dutch. Randal was born in Groningen, and had spent most of his early childhood in Haarlem. His father was a particle physicist, and there were frequent moves, including a two-year stint in Winnipeg, as he followed his work from one experimental nuclear facility to the next.

Now a boyish forty-three, he had lived in California only for the last five years, but had come to think of it as home, or the closest thing to home he'd encountered in the course of a nomadic life. And much of this had to do with the culture of techno-progressivism that had spread outward from its concentrated origins in Silicon Valley and come to encompass the entire Bay Area, with its historically high turnover of radical ideas. It had been a while now, he said, since he'd described his work to someone, only for them to react as though he were making a misjudged joke, or to simply walk off mid-conversation.

Randal was not joking about any of this. For the last thirty years, he had dedicated his life to the ideal of extracting the minds of individuals from the material–flesh, blood, neural tissue–in which they have traditionally been embedded. And this was not an interest he'd happened upon by way of his study of neuroscience. This was an obsession that has shaped his life from the age of thirteen.

The fact that the project of mind uploading, were it to finally succeed, would lead to the effective immortality of the digitally duplicated self was obviously a major area in this whole speculative field, but it wasn't something that especially exercised Randal–not as an end in itself, at any rate. His interest in uploading, he told me, came out of a preoccupation with the limitations of creativity, a precocious awareness of how many things he wished to do and experience, and how little time was allowed for the pursuit of those projects.

"I couldn't optimize problems in my head the way a computer could," he said, taking a neat sip from his beer. "I couldn't work on some problem for a thousand years, or even travel to the next solar system, because I'd be long dead by then. There were so many restrictions, and I realized they all came down to the brain. It was clear to me that the human brain needed enhancement."

In his early teens, Randal began to conceive of the major problem with the human brain in computational terms: the human brain was not, like a computer, readable and rewritable. You couldn't get in there and enhance it, make it run more efficiently, like you could with lines of code. You couldn't just speed up a neuron like you could with a computer processor.

Around this time, he read Arthur C. Clarke's *The City and the Stars*, a novel set in a future a billion years from now, in which the enclosed city of Diaspar is ruled by a superintelligent Central Computer, which creates bodies for the city's posthuman citizens, and stores their minds in its memory banks at the end of their lives, for purposes of future reincarnation. Randal saw nothing in this idea of reducing human beings to data that seemed to him implausible, and felt nothing in himself that prevented him from working to bring it about. His parents encouraged him in this peculiar interest, and the scientific prospect of preserving human minds in hardware became a regular topic of dinnertime conversation.

The emerging discipline of computational neuroscience, which drew its practitioners not from biology but from the fields of mathematics and physics, seemed to offer the most promising approach to the problem of mapping and uploading the mind. It wasn't until he began using the Internet in the mid-1990s, though, that he discovered a loose community of people with an interest in the same area.

As a PhD student in computational neuroscience at McGill University in Montreal, Randal was initially cautious about revealing the underlying motivation for his studies, for fear of being taken for a fantasist or an eccentric.

"I didn't hide it, as such," he said, "but it wasn't like I was walking into labs, telling people I wanted to upload human minds to computers either. I'd work with people on some related area, like the encoding of memory, for instance, with a view to figuring out how that might fit into an overall road map for whole brain emulation."

Having worked for a while at Halcyon Molecular, a Silicon Valley gene-sequencing and nanotechnology start-up funded by Peter Thiel, he decided to stay in the Bay Area and start his own nonprofit aimed at advancing the cause to which he'd long been dedicated. He conceived of Carboncopies as a kind of central gathering point where researchers in various fields—nanotechnology, artificial intelligence, brain imaging, cognitive psychology, biotechnology—crucial to the development of substrate independent minds could meet to share their work and discuss its potential contribution to the cause. Randal described his role within this as an essentially managerial one, although his position was not one with any kind of top-down structural authority.

"I make a lot of phone calls," as he put it. "I don't have postdocs or research assistants. What I have is collaborators, people who feed me information from various sources."

Randal's decision to work outside the academy was rooted in the very reason he began pursuing that work in the first place: an anxious awareness of the small and diminishing store of days that remained to him. If he'd gone the university route, he'd have had to devote most of his time, at least until securing tenure, to projects that were at best tangentially relevant to his central enterprise. The path he had chosen was a difficult one for a scientist, and he lived and worked from one small infusion of private funding to the next. But Silicon Valley's culture of radical techno-optimism had been its own sustaining force for him, and a source of financial backing for a project that took its place within the wildly aspirational ethic of that cultural context. There were people there or thereabouts, wealthy and influential people, for whom a future in which human minds might be uploaded to computers was one to be actively sought—a problem to be solved, disruptively innovated, by the application of money.

One such person was Dmitry Itskov, a thirty-four-year-old Russian tech multimillionaire and founder of the 2045 Initiative, an organization whose stated aim was "to create technologies enabling the transfer of an individual's personality to a more advanced non-biological carrier, and extending life, including to the point of immortality." One of Itskov's projects was the creation of "avatars"—artificial humanoid bodies that would be controlled through brain-machine interface, technologies that would be complementary with uploaded minds. He had funded Randal's work with Carboncopies, and in 2014 they organized a conference together at New York's Lincoln Center called Global Future 2045, aimed, according to its promotional blurb, at the "discussion of a new evolutionary strategy for humanity."

When we spoke, Randal was working with another tech entrepreneur named Bryan Johnson, who had sold his automated payment company to PayPal a couple of years back for $800 million, and who now controlled a venture capital concern called the OS Fund, which, I learned from its website, "invests in entrepreneurs working towards quantum leap discoveries that promise to rewrite the operating systems of life." This language struck me as strange and unsettling in a way that revealed something crucial about the attitude toward human experience that was spreading outward from its Bay Area epicenter—a cluster of software metaphors that had metastasized into a way of thinking about what it meant to be a human being. (Here's how Johnson had put it in a manifesto on the fund's website: "In the same way that computers have operating systems at their core—dictating the way a computer works and serving as a foundation upon which all applications are built—everything in life has an operating system (OS). It is at the OS level that we most frequently experience a quantum leap in progress.")

And it was the same essential metaphor that lay at the heart of Randal's emulation project: the mind as a piece of software, an application running on the platform of flesh. When he used the term "emulation," he was using it explicitly to evoke the sense in which a PC's operating system could be emulated on a Mac, as what he called "platform independent code."

The relevant science for whole brain emulation is, as you'd expect, hideously complicated, and its interpretation deeply ambiguous, but if I can risk a gross oversimplification here, I will say that it is possible to conceive of the idea as something like this: First, you scan the pertinent information in a person's brain—the neurons, the endlessly ramifying connections between them, the information-processing activity of which consciousness is seen as a by-product—through whatever technology, or combination of technologies, becomes feasible first (nanobots, electron microscopy, etc.). That scan then becomes a blueprint for the reconstruction of the subject brain's neural networks, which is then converted into a computational model. Finally, you emulate all of this on a third-party non-flesh-based substrate: some kind of supercomputer, or a humanoid machine designed to reproduce and extend the experience of embodiment—something, perhaps, like Natasha's Primo Posthuman.

The whole point of substrate independence, as Randal pointed out to me whenever I asked him what it would be like to exist outside of a human body—and I asked him many times, in various ways—was that it would be like no one thing, because there would be no one substrate, no one medium of being.

This was the concept transhumanists referred to as "morphological freedom": the liberty to take any bodily form technology permits.

"You can be anything you like," as an article about uploading in *Extropy* magazine put it in the mid-1990s. "You can be big or small; you can be lighter than air, and fly; you can teleport and walk through walls. You can be a lion or an antelope, a frog or a fly, a tree, a pool, the coat of paint on a ceiling."

What really interested me about this idea was not how strange and far-fetched it seemed (though it ticked those boxes resolutely enough), but rather how fundamentally identifiable it was, how universal. When I was talking to Randal, I was mostly trying to get to grips with the feasibility of the project, and with what it was he envisioned as a desirable outcome. But then we would part company—I would hang up the call, or I would take my leave and start walking toward the nearest BART station—and I would find myself feeling strangely affected by the whole project, strangely moved.

Because there was something, in the end, paradoxically and definitively human in this desire for liberation from human form. I found myself thinking often of W. B. Yeats's "Sailing to Byzantium," in which the aging poet writes of his burning to be free of the weakening body, the sickening heart—to abandon the "dying animal" for the man-made and immortal form of a mechanical bird. "Once out of nature," he writes, "I shall never take/My bodily form from any natural thing/But such a form as Grecian goldsmiths make."

Yeats, clearly, was not writing about the future so much as an idealized phantasm of the ancient world. But the two things have never been clearly separated in our minds, in our cultural imaginations. All utopian futures are, in one way or another, revisionist readings of a mythical past. Yeats's fantasy here is the fantasy of being an archaic automaton invested with an incorruptible soul, a mechanical bird singing eternally. He was writing about the terror of aging and bodily decline, about the yearning for immortality. He was asking the "sages" to emerge from a "holy fire" and to gather him "into the artifice of eternity." He *was* dreaming of a future: an impossible future in which he would not die. He was dreaming, I came to feel, of a Singularity. He was singing of what was past, and passing, and to come.

In May 2007, Randal was one of thirteen participants at a workshop on mind uploading held at the Future of Humanity Institute. The event resulted in the publication of a technical report, coauthored by Anders Sandberg and Nick Bostrom, entitled "Whole Brain Emulation: A Roadmap." The report began with the statement that mind uploading, though still a remote prospect, was nonetheless theoretically achievable through the development of technologies already in existence.

A criticism commonly raised against the idea of simulating minds in software is that we don't understand nearly enough about how consciousness works to even know where to start reproducing it. The report countered this criticism by claiming that, as with computers, it wasn't necessary to comprehend a whole system in order to emulate it; what was needed was a database containing all the relevant information about the brain in question, and the dynamic factors that determine changes in its state from moment to moment. What was needed, in other words, was not an understanding of the information, but merely the information per se, the raw data of the person.

A major requirement for the harvesting of this raw data, they wrote, was "the ability to physically scan brains in order to acquire the necessary information." A development that appeared especially promising in this regard was something called 3D microscopy, a technology for producing extremely high-resolution three-dimensional scans of brains.

Another of the workshop's invited participants was a man named Todd Huffman, the CEO of a San Francisco company called 3Scan, which happened to be pioneering exactly this technology. Todd was among the collaborators Randal had mentioned–one of the people who kept him regularly updated about their work, and its relevance to the overall project of uploading.

Although one of 3Scan's initial sources of start-up funding was Peter Thiel–a man who, although he did not explicitly identify with the transhumanist movement, was famously invested in the cause of vastly extending human life spans, in particular his own–it was not a company that had any explicit designs on the brain uploading market. (And the major reason for this was that such a market was nowhere near existing.) It promoted its technology as a tool for the diagnosis and analysis of cell pathologies, as a medical device. But when I met with Todd at 3Scan's offices in Mission Bay, he was open about the extent to which his work was motivated by a long-standing preoccupation with translating individual human minds into computable code. He was not, he said, interested in standing on the sidelines and waiting for the Singularity to just happen, by sheer force of some quasi-mystical historical determinism.

"You know what they say," said Todd. "The best way to predict the future is to create it."

He was a fully paid-up transhumanist, Todd: he was a member of Alcor, and had an implant in the tip of his left ring finger that, by means of a mild vibration, allowed him to sense the presence of electromagnetic fields. Visually, he was a cut-up composite of two or three different guys: vigorous rustic beard, plumage of pink hair, Birkenstocks, black-painted toenails.

The people who worked for him, he said, knew of his long-term interest in whole brain emulation, but it was not something that drove the company in its day-to-day dealings. It just so happened that the sort of technology that would ultimately prove useful for scanning human brains for emulation was, he said, useful right now for more immediate projects like analyzing pathologies for cancer research.

"The way I look at it," he said, "is that mind uploading is not driving industry, but industry is driving mind uploading. There are a lot of industries that have nothing to do with mind uploading, but that are driving the development of technologies that will be used for uploading. Like the semiconductor industry. That space has developed techniques for very fine-grained milling and measurement, and also the kind of electron microscopes that turn out to be very useful for doing high-resolution 3D reconstructions of neurons."

The moon-shot ethos of Silicon Valley was such that Todd never felt uncomfortable discussing his interest in brain uploading, but neither was it something that tended to come up in business meetings. There was a very small community, he said, of people who were thinking seriously about this stuff at a high scientific level, and an even smaller number of people who were working on it.

"I know people who are doing work on uploading," he said, "and doing it in secret, because they're afraid of being ostracized within their scientific communities, or of being passed over for funding or tenure or promotions. I don't have that; I work for myself, so no one is going to throw me out of the building."

He walked me around the lab, cracking his knuckles intermittently as we moved among the bewildering assemblages of optics and digitizing devices, the fine slices of rodent brains preserved in glass like ostentatious servings of neural carpaccio. These slices had been imaged and digitized using the 3D microscopes, for detailed databasing–of neuron placement, dimensions and arrangements of axons, dendrites, synapses.

Looking at these brain slices, I understood that, even if a greatly scaled-up version of this scanning technology eventually made it possible to perform whole brain emulation, it would be impossible to emulate the brain of an animal without killing that animal–or at least killing the original, embodied version. This was an acknowledged problem among advocates of emulation, and the idea of nanotechnology—technology on a scale sufficiently minuscule for the manipulation of individual molecules and atoms—was an area that offered some hope. "We can imagine," writes Murray Shanahan, a professor of cognitive robotics at Imperial College London, "creating swarms of nano-scale robots capable of swimming freely in the brain's network of blood vessels, each one then attaching itself like a limpet to the membrane of a neuron or close to a synapse." (Randal, for his part, spoke enthusiastically of something called "neural dust," a technology that was being developed at U.C. Berkeley which would allow the application of infinitesimally tiny wireless probes to neurons, allowing the extraction of data without causing any damage. "It'd be like taking an aspirin," he said.)

I began to think of these brain slices as illustrating the strange triangulations of the relationship between humans and nature and technology. Here was a sliver of an animal's central nervous system, pressed and mounted in glass in order to make its contents readable by a machine. What did it mean to do this to a brain—to an animal brain, a human brain? What would it mean to make consciousness readable, to translate the inscrutable code of nature into the vulgate of machines? What would it mean to extract information from such a substrate, to transpose it to some other medium? Would the information mean anything at all outside the context of its origins?

I felt suddenly the extreme strangeness of this notion of ourselves as essentially information, as contained in some substrate that was not what we were, but merely the medium for our intelligence—as though our bodies might be categorized, along with the glass slides in which these brain slices were preserved, as mere casing. A certain kind of extreme positivist view of human existence insists that what we are is *intelligence;* and intelligence, as well as referring to the application of skills and knowledge, also means information that is gathered, extracted, filed.

"Most of the complexity of a human neuron," writes Ray Kurzweil, "is devoted to maintaining its life-support functions, not its information processing capabilities. Ultimately, we will be able to port our mental processes to a more suitable computational substrate. Then our minds won't have to stay so small."

At the root of this concept of whole brain emulation, and of transhumanism itself as a movement or an ideology or a theory, was, I realized, the sense of ourselves as trapped in the wrong sort of stuff, constrained by the material of our presence in the world. To talk of achieving a "more suitable computational substrate" only made sense if you thought of yourself as a computer to begin with.

In philosophy of mind, the notion that the brain is essentially a system for the processing of information, and that in this it therefore resembles a computer, is known as computationalism. As an idea, it predates the digital era. In his 1655 work *De Corpore,* for instance, Thomas Hobbes wrote, "By reasoning, I understand computation. And to compute is to collect the sum of many things added together at the same time, or to know the remainder when one thing has been taken from another. To reason therefore is the same as to add or subtract."

And there has always been a kind of feedback loop between the idea of the mind as a machine, and the idea of machines with minds. "I believe that by the end of the century," wrote Alan Turing in 1950, "one will be able to speak of machines thinking without expecting to be contradicted."

As machines have grown in sophistication, and as artificial intelligence has come to occupy the imaginations of increasing numbers of computer scientists, the idea that the functions of the human mind might be simulated by computer algorithms has gained more and more momentum. In 2013, the EU invested over a billion euros of public funding in a venture called the Human Brain Project. The project, based in Switzerland and directed by the neuroscientist Henry

Markram, was set up to create a working model of a human brain and, within ten years, to simulate it on a supercomputer using artificial neural networks.

Not long after I left San Francisco, I traveled to Switzerland to attend something called the Brain Forum, an extravagantly fancy conference on neuroscience and technology at the University of Lausanne, where the Human Brain Project is based. One of the people I met there was a Brazilian named Miguel Nicolelis, a professor at Duke University. Nicolelis is one of the world's foremost neuroscientists, and a pioneer in the field of brain-machine interface technology, whereby robotic prostheses are controlled by the neuronal activity of human beings. (Randal had referred to this technology several times during our discussions.)

Nicolelis was a fulsomely bearded man with an impish manner about him; the Nikes he wore with his suit seemed less an affectation than an insistence on comfort over convention. He was in Lausanne to give a talk about a brain-controlled robotic exoskeleton he had developed, which had allowed a quadriplegic man to kick the first ball during the 2014 World Cup opening ceremony in São Paulo.

Given the frequency with which his own work was cited by transhumanists, I was curious to find out what Nicolelis thought about the prospect of mind uploading. He did not, it turned out, think much of it. The whole idea of simulating a human mind in any kind of computational platform was fundamentally at odds, he said, with the dynamic nature of brain activity, of what we think of as the mind. It was for this same reason, he said, that the Human Brain Project was utterly ill conceived.

"The mind is much more than information," he said. "It is much more than data. That's the reason you can't use a computer to find out how the brain works, what is going on in there. The brain is simply not computable. It cannot be simulated."

Brains, like many other naturally occurring phenomena, processed information; but this didn't mean, for Nicolelis, that such processing could be rendered algorithmically and run on a computer. The central nervous system of a human being had less in common with a laptop than it did with other naturally occurring complex systems like schools of fish or flocks of birds—or, indeed, stock markets—where elements interact and coalesce to form a single entity whose movements were inherently unpredictable. As he put it in *The Relativistic Brain,* a book coauthored with the mathematician Ronald Cicurel, brains constantly reorganize themselves, both physically and functionally, as a result of actual experience: "Information processed by the brain is used to reconfigure its structure and function, creating a perpetual recursive integration between information and brain matter. . . . The very characteristics that define a complex adaptive system are the ones that undermine our capacity to accurately predict or simulate its dynamic behavior."

Nicolelis's skepticism about the computability of the brain put him in a minority at the Brain Forum. No one was talking about anything as remote and abstract as brain uploading, but almost every sentence I heard reinforced the consensus that the brain could be translated into data. The underlying message of the conference as a whole seemed to be that scientists were still almost entirely ignorant about how the brain did what it did, but the scanning of brains and the building of vast dynamic models was absolutely necessary if we were to begin learning the first thing about what was going on inside our heads.

The following day, I met a neuroengineer named Ed Boyden. Boyden, a bearded and bespectacled and serenely exuberant American in his mid-thirties, led the Synthetic Neurobiology research group at MIT Media Lab. His work involved building tools for mapping and controlling and observing the brain, and using them to figure out how the thing actually works. He had gained considerable fame in recent years for his role in the creation of optogenetics, a neuromodulation technique whereby individual neurons in the brains of living animals could be switched on and off by the application of directed light photons.

Randal had mentioned his name on several occasions during our discussions—both as someone broadly supportive of whole brain emulation and whose work was of significant relevance to that project–and Boyden had been a speaker at the Global Future 2045 event in New York the previous year.

It was Boyden's belief, he told me, that it would eventually be possible to build neuroprosthetic replacements for brain parts–which, if you take the Ship of Theseus view of things, is essentially the same as believing that whole brain emulation is possible.

"Our goal is to solve the brain," he said. He was referring here to the ultimate goal of neuroscience, which was to understand how the brain did what it did, how its billions of neurons, and the trillions of connections between them, organized themselves in such a way as to produce specific phenomena of consciousness. I was struck by the mathematical implications of the term *solve,* as though the brain could, in the end, be *worked out* like an equation or a crossword puzzle.

"To solve the brain," he said, "you have to be able to simulate it in a computer. We're working very hard on ways to map the brain, using connectomics. But I would argue that connections are not enough. To understand how information is being processed, what you really need are all the molecules in the brain. And I think a reasonable goal at this point would be to simulate a small organism, but to do this you need a way to map a 3D object such as the brain with nanoscale precision."

Boyden's team at MIT, as it happened, had recently developed just such a radical tool. It was called expansion microscopy, and it involved physically inflating samples of brain tissue using a polymer most commonly found in baby's diapers. The polymer allowed for a scale blowup of the tissue, which kept all the proportions and connections in place, and facilitated a radically increased level of detail in mapping.

Boyden took out his laptop and showed me some 3D images of brain tissue samples that had been made using the technique.

"So what's the ultimate aim of this?" I asked him.

"Well, I think it'd be great if we could actually localize and identify all the key proteins and molecules in the brain circuit. And then you could potentially make a simulation, and model what's going on in the brain."

"When you say simulation, what are you talking about? Are you talking about a functioning, conscious mind?"

Boyden paused for a moment and, in a quietly rhetorical flourish, confessed to not really understanding what the term "consciousness" meant—at least not precisely enough to answer my question.

"The problem with consciousness as a word," he said, "is we have no way of judging whether it's there or not. There's not like a test that you can run and if it scores ten or higher, then that's consciousness. So it's hard to know whether a simulation would be conscious per se."

He gestured toward the laptop in front of him on the table, in the conference venue's vast empty banquet room where we'd come to talk, and he said that in order to understand a computer, it was not enough to understand the wiring, you needed to understand the dynamics.

"There are five hundred million of these laptops on the planet," he said, "and they all have the same static wiring, but right now, at this moment, they're all doing different things dynamically. So you need to understand the dynamics, not just what's in there in terms of wiring and microchips and so on."

He clicked around for a few seconds on his trackpad, and brought up an image of a worm, animated with twinkling colored spots of light. This was the *C. elegans* nematode, a transparent roundworm about a millimeter in length, much favored by neurologists for its manageably tiny number of neurons (302). This worm was the first multicellular organism ever to have its genome sequenced, and is to date the only creature to have its connectome fully mapped.

"So this is the first attempt to image all the neural activity in a whole organism," he said, "at a rate that's fast enough to capture the activation of all those neurons. And so if we can capture the connectivity and molecules in a circuit, and if we can watch what's happening in real time, then we can try to really see whether the simulated dynamics recapitulates the empirical observation."

"At which point you'll what? Be able to translate this worm's neural activity into code? Into a computable form?"

"Yes," said Boyden. "That's the hope."

I felt that he was holding back from telling me he believed that whole brain emulations would at some point become a reality, but it was clear that he felt the principle to be sound, in a way that Nicolelis did not. And what he was telling me, ultimately, was that whether or not it led to it in the end, and whether or not it was his own ultimate goal, the kind of research that was necessary for the achievement of whole brain emulation was precisely the kind of research he himself was doing at MIT.

This was all clearly a very long way from where Randal wanted to get to, a very long way from his mind, or mine, or yours, on a laptop screen, with its hundred billion firing neurons glimmering with the light of purified consciousness. But it was an illustration of a principle, a statement of a possibility: an indicator that this thing that Randal wanted to do was not entirely crazy, or not, at least, entirely outside of the bounds of the thinkable.

In my first couple of conversations with Randal, my questions tended to focus on the technical aspects of whole brain emulation—on the means by which it might be achieved, and on the overall feasibility of the project. This was useful insofar as it confirmed for me that Randal at least knew what he was talking about, and that he was not insane, but this should not be taken to imply that I myself understood these matters in anything but the most rudimentary fashion.

One evening, we were sitting outside a combination bar/laundromat/stand-up comedy venue on Folsom Street—a place with the fortuitous name of Brainwash—when I confessed to Randal that the idea of having my mind uploaded to some technological substrate was deeply unappealing to me, horrifying even. The effects of technology on my own life, even now, were something about which I was profoundly ambivalent; for all I had gained in convenience and "connectedness," I was increasingly aware of the extent to which my movements in the world were mediated and circumscribed by corporations whose only real interest was in reducing the lives of human beings to data, as a means to further reducing us to profit. The "content" we consumed, the people with whom we had romantic encounters, the news we read about the outside world: all these movements were coming increasingly under the influence of unseen algorithms, the creations of these corporations—whose complicity with government, moreover, had come to seem like the great submerged narrative of our time. Given the world we were now living in, where the fragile liberal ideal of the autonomous self was already receding like a half-remembered dream into the doubtful haze of history, wouldn't a radical fusion of ourselves with technology amount, in the end, to a final capitulation of the very idea of personhood?

Randal nodded again, and took a sip of his beer.

"Hearing you say that," he said, "makes it clear that there's a major hurdle there for people. I'm more comfortable than you are with the idea, but that's because I've been exposed to it for so long that I've just got used to it."

The most persistently troubling philosophical question raised by all of this is also the most basic: Would it be me? If the incalculable complexity of my neural pathways and processes could somehow be mapped and emulated and run on a platform other than the 3.3 lbs of gelatinous nervous tissue contained inside my skull, in what sense would that reproduction or simulation be "me"? Even if you allow that the upload is conscious, and that the way in which that consciousness presents itself is indistinguishable from the way I present myself, does that make it me? If the upload believes itself to be me, is that enough? (Is it enough that I believe myself to be me right now, and does that even mean anything at all?)

I had a very strong feeling—an instinctual burst of subcortical signals—that there was no distinction between "me" and my body, that I could never exist independently of the substrate on which I operated because the self was the substrate, and the substrate was the self.

The idea of whole brain emulation—which was, in effect, the liberation from matter, from the physical world—seemed to me an extreme example of the way in which science, or the belief in scientific progress, was replacing religion as the vector of deep cultural desires and delusions.

Beneath the talk of future technologies, I could hear the murmur of ancient ideas. We were talking about the transmigration of souls, eternal return, reincarnation. Nothing is ever new. Nothing ever truly dies, but is reborn in a new form, a new language, a new substrate.

We were talking about immortality: the extraction of the essence of the person from the decaying structure of the body, the same basic deal humanity had been dreaming of closing since at least as far back as Gilgamesh. Transhumanism is sometimes framed as a contemporary resurgence of the Gnostic heresies, as a quasi-scientific reimagining of a very ancient religious idea. ("At present," as the political philosopher John Gray puts it, "Gnosticism is the faith of people who believe themselves to be machines.") The adherents of this early Christian heretical sect held that the material world, and the material bodies with which human beings negotiated that world, were the creation not of God but of an evil second-order deity they called the demiurge. For the Gnostics, we humans were divine spirits trapped in a flesh that was the very material of evil. In his book *Primitive Christianity,* Rudolf Bultmann quotes a passage from a Gnostic text outlining what must be done in order to ascend to the realm of divine light:

> First thou must rend the
> garment that now thou
> wearest, the attire of igno-
> rance, the bulwark of evil,

> the bond of corruption,
> the dark prison, the living
> death, the sense-endowed
> corpse, the grave thou
> bearest about with thee,
> the grave, which thou ca
> -rriest around with thee, the
> thievish companion who
> hateth thee in loving thee,
> and envieth thee in hating
> thee...

It was only through the achievement of higher refinements of knowledge that an elect few–the Gnostics themselves, initiates of divine information–could escape from the evil of embodiment into the rarefied truth of pure spirit. The Jesus of the Gnostic Apocrypha is contemptuous of the body in a way that is much more explicit and unambiguous than anything in canonical Scripture. In the Gnostic Gospel of Thomas, He is quoted as saying that "If spirit came into being because of the body, it is a wonder of wonders. Indeed, I am amazed at how this great wealth has made its home in this poverty."

These beliefs, as Elaine Pagels puts it in her book *The Gnostic Gospels*, "stood close to the Greek philosophic tradition (and, for that matter, to Hindu and Buddhist tradition) that regards the human spirit as residing 'in' a body–as if the actual person were some sort of disembodied being who uses the body as an instrument but does not identify with it." For the Gnostics, the only redemption would come in the form of liberation from that body. And a technological version of this liberation seemed to me to be what whole brain emulation was ultimately all about.

This techno-dualistic account of ourselves, as software running on the hardware of our bodies, had grown out of an immemorial human propensity to identify ourselves with, and explain ourselves through, our most advanced machines. In a paper called "Brain Metaphor and Brain Theory," the computer scientist John G. Daugman outlines the history of this tendency. Just as the water technologies of antiquity (pumps, fountains, water-based clocks) gave rise to the Greek and Roman languages of pneuma and the humors; and just as the presiding metaphor for human life during the Renaissance was clockwork; and just as in the wake of the Industrial Revolution, with its steam engines and pressurized energies, Freud brought these forces to bear on our conception of the unconscious, there was now a vision of the minds of humans as devices for the storing and processing of data, as neural code running on the wetware of the central nervous system.

If we are anything at all, in this view, what we are is information, and information has become an unbodied abstraction now, and so the material through which that information is transmitted is of secondary importance to its content, which can be endlessly transferred, duplicated, preserved. ("When information loses its body," writes the literary critic N. Katherine Hayles, "equating humans and computers is especially easy, for the materiality in which the thinking mind is instantiated appears incidental to its essential nature.")

A strange paradox lies at the heart of the idea of simulation: it arises out of an absolute materialism, out of a sense of the mind as an emergent property of the interactions between physical things, and yet it manifests as a conviction that mind and matter are separate, or separable. Which is to say that it manifests as a new form of dualism, even a kind of mysticism.

The more time I spent with Randal, the more preoccupied I became with finding out what exactly it was that he envisioned when he thought about the eventual achievement of his project. What would be the experience of an uploaded version of the self? What did he imagine it would feel like to be a digital ghost, a consciousness untethered to any physical thing?

His answers varied whenever I asked him about this, and he was open about the fact that he didn't have any one clear picture. It would depend, he told me, on the substrate; it would depend on the material of being. Sometimes he told me that there would always be a material presence, some version of flesh and blood, and then sometimes he would invoke the idea of virtual selves, of embodied presences in virtual worlds.

"I often think," he said, "that it might be like the experience of a person who is, say, really good at kayaking, who feels like the kayak is physically an extension of his lower body, and it just totally feels natural. So maybe it wouldn't be that much of a shock to the system to be uploaded, because we already exist in this prosthetic relationship to the physical world anyway, where so many things are experienced as extensions of our bodies."

I realized, at this point, that I was holding my phone in my hand; I placed it on the table, and we both smiled.

I mentioned to Randal some concerns I had about potential consequences of his project. I was already troubled enough, I said, about the extent to which modern lives had been converted into code, into highly transferable and marketable stockpiles of personal information. Our every engagement with technology created an increasingly detailed portrait of our consumer selves, which was the only version of the self that mattered to the makers of these technologies. How much worse would it be if we existed purely as information? Would consciousness itself become a form of cognitive clickbait? Even now, I said, I was already imagining some terrifying extrapolation of native advertising, whereby my inclination to order another Sierra Nevada arose not out of any self-contained nexus of desire and volition, but from some clever piece of code that had been frictionlessly insinuated into the direct-marketing platform of my consciousness.

What if the immortalization procedure, the emulation and the upload, wound up being so expensive that only the extremely wealthy could afford the ad-free premium subscription, and the rest of us losers had to put up with subsidizing our continued existence through periodic exposure to thoughts or emotions or desires imposed from above, by some external commercial source, in some hellish sponsored content partnership of the self?

Randal did not disagree that such a situation would be undesirable. None of it, though, was directly relevant to his immediate project here, which was to solve the basic problem of human embodiment, rather than to head off at the pass any unintended consequences thereof.

"Plus," he said, "it's not like that kind of influence is exactly unique to software. It can be done with biological brains. By means of advertising, say. Or by means of chemicals. It's not like you wanting another beer doesn't have anything to do with the alcohol you've already consumed. It's not like your desires are entirely independent of outside influence."

I took a long drink of my beer, resolving as I did so to forgo ordering a second. A stench of weed had settled heavily on the warm evening, like a dank fog drifting in from the bay, and the air itself seemed in the grip of a teeming, paranoid high. Just a few feet from where we were sitting, on the corner of Folsom and Langton Streets, a young homeless man lay curled in the fetal position by a lamppost; he'd been keeping up a low, muttering monologue the entire time we'd been there, and as I put down my beer and thought about the things Randal had been saying, the man, whose face I could not see, let out a high, hysterical series of rapid-fire titters. I found myself thinking of a line from Nietzsche's *The Gay Science*, about how strange, how uncannily wrong-natured, we must seem to other animals: "I fear that the animals see the human being as a being like themselves who in a most dangerous manner has lost its animal common sense—as the insane animal, as the laughing animal, the weeping animal, the unhappy animal."

Perhaps the reason for our being insane animals is precisely our inability to accept ourselves as animals, to accept the fact that we will die animal deaths. And why should we accept it? It's a fact not to be borne, an inadmissible reality. You would think that we'd be beyond this; you would think that, by now, we'd do better than just succumbing to nature's final dumb imperative. Our existence, and its attendant neurosis, is defined by a seemingly irresolvable contradiction: that we are outside of nature, beyond it and above it like minor deities, and yet always helplessly within it, forever defined and circumscribed by its blind and implacable authority.

I felt that I was catching a glimpse of the absurdity that lay beneath everything we thought of as the world—beneath reason, and science, and the idea of human progress. Everything seemed suddenly, giddyingly revealed as bizarre and self-evidently preposterous: the scientist talking about liberating men and women from the captivity of flesh, the malfunctioning mechanism of the homeless man in a heap on a San Francisco sidewalk muttering his madness and misery into a void, the writer deluding himself with thoughts of seeing into the heart of things, and making a note to write something about the tittering derelict, the smell of weed, Nietzsche's mad animals.

In the weeks and months after I returned from San Francisco, I thought obsessively about the idea of whole brain emulation. I would be taking a break from work and walking to a coffee shop, and a car would drive past me a little too fast, and I would have an image of that car mounting the footpath at speed and plowing into me; I would imagine what the impact might do to my body, and I would find myself thinking of Randal and his project of separating the self from the substrate. If I felt tired or physically vulnerable, I (or my mind, or my brain) was especially likely to return to Randal, or to the worm neurons I'd seen shimmering on Ed Boyden's laptop in Lausanne, or to the slices of preserved mouse brain I'd seen at 3Scan's laboratory in Mission Bay.

One morning, some months after I returned from that trip to San Francisco, I was at home in Dublin, suffering from both a head cold and a hangover—the severity of which latter condition seemed to me out of all reasonable proportion to the moderate amount I'd drunk the previous night. I lay there, idly considering hauling myself out of bed to join my wife and my son, who were in his bedroom next door enjoying a raucous game of Buckaroo. I realized that these conditions (head cold, hangover) had imposed upon me a regime of mild bodily estrangement. As often happens when I'm feeling under the weather, I had a sense of myself as an irreducibly biological thing, an assemblage of flesh and blood and gristle. I felt myself to be an organism with blocked nasal passages, a bacteria-ravaged throat, a sorrowful ache deep within its skull, its *cephalon.* I was aware of my substrate, in short, because my substrate felt like shit.

And I was gripped by a sudden curiosity as to what, precisely, that substrate consisted of–as to what I myself happened, technically speaking, to be. I reached across for the phone on my nightstand, and entered into Google the words "What is the human..." The first three autocomplete suggestions offered "What is The Human Centipede about"; and then: "What is the human body made of"; and then: "What is the human condition." It was the second question I wanted answered at this particular time, as perhaps a back door into the third.

It turned out that I was 65 percent oxygen–which is to say that I was mostly air, mostly nothing. After that, I was composed of diminishing quantities of carbon and hydrogen, of calcium and sulfur and chlorine and so on down the elemental table. I was also mildly surprised to learn that, like the iPhone I was extracting this information from, I also contained trace elements of copper and iron and silicon.

What a piece of work is a man, I thought; what a quintessence of dust.

Some minutes later, my wife entered the bedroom on her hands and knees, our son on her back, gripping the collar of her shirt tight in his little fists. She was making clip-clop noises as she crawled forward, and he was laughing giddily, and shouting "Don't buck! Don't buck!"

With a loud neighing sound, she arched her back and sent him tumbling gently into a row of shoes by the wall, and he screamed in delighted outrage, before climbing up again.

None of this, I felt, could be rendered in code. None of this, I felt, could be run on any other substrate. Their beauty was bodily, in the most profound sense, in the saddest and most wonderful sense.

I never loved my wife and our little boy more, I realized, than when I thought of them as mammals. I dragged myself, my animal body, out of bed to join them.

A Short Note on the Singularity

THERE IS NO one accepted version of the Technological Singularity. It is a light gleaming above Silicon Valley's horizon, appearing now as religious prophecy, now as technological fate. There is no end to the riches the faithful claim it will generate, no end to what can be said of them. In the broadest sense, the term refers to a time to come in which machine intelligence greatly surpasses that of its human originators, and biological life is subsumed by technology. It is, in its way, an extreme expression of techno-progressivism, the belief that the universal application of technology will solve the world's most intractable problems.

The idea has been around in some form for at least half a century. In his 1958 obituary for the physicist John von Neumann, with whom he had worked on the Manhattan Project, Stanislaw Ulam wrote about a conversation they once had about "the ever accelerating progress of technology and changes in the mode of human life, which gives the appearance of approaching some essential singularity in the history of the race beyond which human affairs, as we know them, could not continue."

The first substantial statement of the concept of a Technological Singularity is usually attributed to the mathematician and science fiction writer Vernor Vinge. In an essay called "The Coming Technological Singularity: How to Survive in the Post-human Era," first delivered as a paper at a 1993 conference organized by NASA, Vinge claimed that "within thirty years, we will have the technological means to create superhuman intelligence. Shortly thereafter, the human era will be ended." Vinge is ambivalent about the consequences of this great transcendence: it could mean the end of all our problems, or the annihilation of our species, but that it is coming is not seriously in doubt. As with much of this style of techno-millenarian thinking, Vinge's prophesying is characterized by a strange sort of historical determinism: there can be no preventing the Singularity, he writes, because its coming is an inevitable consequence of our natural competitiveness, and the inherent possibilities of technology. "And yet," he writes, "we are the initiators."

The closest thing to a canonical Singularity is the version that appears in the popular writings of Ray Kurzweil. Kurzweil is an inventor of many ingenious devices—the flatbed scanner, the print-to-speech reading machine for the blind—and the cofounder, with Stevie Wonder, of Kurzweil Music Systems, whose synthesizers are used by such diverse acts as Scott Walker and New Order and "Weird Al" Yankovic. As a writer, he is a controversial figure, a business-casual mystic whose arcane projections chart the furthest reaches of techno-utopian speculation. But he is by no means a marginal presence in the tech world; he is, rather, a tutelary spirit of Silicon Valley—a status that was more or less formalized in 2012, when he was brought in as director of engineering at Google, to act as thought-leader-in-chief for the company's pursuit of machine learning.

Kurzweil's Singularity is a wildly multifarious vision of technological abundance, a feverishly detailed teleology in which all of history converges toward an apotheosis of pure mind. "How do we contemplate the Singularity?" he asks in the early pages of his 2005 bestseller, *The Singularity Is Near: When Humans Transcend Biology*. "As with the Sun, it's hard to look at directly; it's better to squint at it out of the corners of our eyes." He is, however, obligingly forthcoming with certain details: the Singularity, for one thing, is informally penciled in for the year 2045 or thereabouts. (Kurzweil, who famously consumes a prodigious daily banquet of dietary supplements and vitamin pills—and who, for that matter, markets his own personal brand of death-forestalling elixirs and capsules—is confident that, at age ninety-seven, he'll still be around.)

As a diviner of the technological future, Kurzweil's principal tool is what he calls "the law of accelerating returns." Technology progresses, in this view, along the same lines as a financial investment with compound interest, which is to say exponentially. Our current technology is the foundation from which we develop future technology, and so the more sophisticated technology becomes, the greater the rate at which it improves. (The best-known example of this phenomenon is an observation, first made in the 1950s by Intel cofounder Gordon Moore and which came to be known as Moore's Law, stating that the number of transistors that can be fit on a single microchip doubles roughly once every eighteen months.) The process of Darwinian evolution is, for Kurzweil, itself a process of exponential growth, and one that is explicitly represented as growing *toward* a desirable end. Evolution is not a blind and chaotic fumbling, a random generation of horrors and wonders, but rather a *system*, "a process of creating patterns of increasing order." Evolution, in other words, is an advancement toward the perfect order and regulation of the machine. And it is this evolution of patterns—a logical progression in which "each stage or epoch uses the information-processing methods of the previous epoch to create the next"—that constitutes, for Kurzweil, "the ultimate story of our world."

The picture Kurzweil paints of the future is one in which technology continues to get smaller and more powerful, until such time as its accelerating evolution becomes the primary agent of our own evolution as a species. We will no longer carry computers around with us, he reveals, but rather take them into our bodies—into our brains and our bloodstreams—changing thereby the nature of the human experience. In the very near future (i.e., hopefully within the lifetime of Kurzweil himself), this will be not merely possible but necessary, given the unsatisfactory computing power of even the most efficient human brain.

Kurzweil's vision of the future might be an attractive one if you already accept the mechanistic view of the human being—if you agree with AI pioneer Marvin Minsky that the brain "happens to be a meat machine." Why would we, or our meat machines, not choose to upgrade to some higher degree of functionality? If we understand a machine to be an apparatus constructed for the performance of a particular task, then our task as machines is, surely, to think, to *compute,* at the highest level possible. In this instrumentalist view of human life, it is more or less our duty—or at least pretty much the whole point of our existing in the first place—to increase our computational firepower, and to ensure that, as machines, we run as efficiently as possible for as long as possible.

"Our version 1.0 biological bodies," writes Kurzweil, are "frail and subject to a myriad of failure modes, not to mention the cumbersome maintenance rituals they require. While human intelligence is sometimes capable of soaring in its creativity and expressiveness, much human thought is derivative, petty, and circumscribed." This, we are assured, will no longer be the case once the Singularity kicks in: we will no longer be helpless and primitive creatures, meat machines restricted in our thoughts and actions by the flesh that is our current substrate. "The Singularity," he writes, "will allow us to transcend these limitations of our biological bodies and brains. We will gain power over our fates. Our mortality will be in our own hands. We will be able to live as long as we want (a subtly different statement from saying we will live forever). We will fully understand human thinking and will vastly extend and expand its reach. By the end of this century, the nonbiological portion of our intelligence will be trillions of trillions of times more powerful than unaided human intelligence."

We will, in other words, finally escape the fallen condition of our humanity, finally become unfleshed; we will be restored to a prelapsarian state of wholeness, a final union in which technology will take the place of the Abrahamic God. "The Singularity," writes Kurzweil, "will represent the culmination of the merger of our biological thinking and existence with our technology, resulting in a world that is still human but that transcends our biological roots. There will be no distinction, post-Singularity, between human and machine or between physical and virtual reality." To the charge that such a merger would obliterate our humanity, Kurzweil counters that the Singularity is in fact a final achievement of the human project, an ultimate vindication of the very quality that has always defined and distinguished us as a species—our constant yearning for a transcendence of our physical and mental limitations.

In *The City of God,* Saint Augustine conjures a state of "universal knowledge" far beyond anything we can now imagine, which will be the preserve of those blessed by God's grace. "Think how great, how beautiful," he writes, "how certain, how unerring, how easily acquired this knowledge will be. And what a body, too, we shall have, a body utterly subject to our spirit and one so kept alive by spirit that there will be no need of any other food."

In Kurzweil's prophecy, the messianic role is occupied by intelligence. Although he ascribes mystical significance to the term, his definition is straightforward enough: he takes intelligence essentially to mean computation—the algorithmic machinery that is brought to bear on the raw informational stuff of creation. And in this messianic vision, machine intelligence will come to redeem the universe of its incalculable stupidity.

He takes a goal-oriented approach to cosmology, imposing upon the universe itself a kind of corporate project-management structure, composed of a series of key deliverables across deep time. In the last of what he calls the "Six Epochs of Evolution," after the great fusion of humanity and AI, intelligence "will begin to saturate the matter and energy in its midst." This will be achieved, writes Kurzweil, "by reorganizing matter and energy to provide an optimal level of computation... to spread out from its origins on Earth." Through careful husbandry, the infinite emptiness of the universe—after some fourteen billion years of just sitting around uselessly succumbing to the inexorable force of entropy—will finally be turned to account as a vast data-processing mechanism.

In one of the stranger scenes in *Transcendent Man*, a 2009 documentary about Kurzweil's life and work, we encounter the film's subject standing on a beach at sunset, his inscrutable gaze fixed upon the serene expanse of the Pacific. In the scene directly preceding, we have just heard him speak movingly about the last conversation he had with his father before his death, and so when we hear the voice of the director asking Kurzweil what it is he is thinking about as he stares out to sea, we might reasonably expect him to say that he is thinking of mortality—if not of his own, at least the mortality of those not fortunate enough to live long enough to live forever. Kurzweil pauses a long moment, and the camera begins to move about him in slow ceremonial circles.

"Well," he says, "I was thinking about how much computation is represented by the ocean. I mean, it's all these water molecules interacting with each other. That's computation. It's quite beautiful. I've always found it very soothing. And that's really what computation is all about. To capture these transcendent moments of our consciousness."

Facing out onto the terminal vastness of the Pacific, his hair slightly ruffled by the breeze, Kurzweil appears an oracular figure, a conduit for the mysteries of technology, a prophet of a world to come, in which an infinite and instrumental intelligence will finally release us from the burden of our humanity.

He looks at the ocean, and sees a vast and intricate device which is nothing but information, nothing but the raw material for intelligence. The water, its fluctuations in temperature, its swarming profusion of organisms, its rhythmic advances and retreats: all of it an immense calculus, a code. The sea, like a certain way of thinking about thought itself, is a patterned manipulation of propositions. And in this moment a kind of computational pantheism reveals itself, a reverence for nature as an expression of a universal machine, an algorithmic immanence.

Talkin' AI Existential Risk Blues

EVEN IF IT were possible to put aside for a moment the considerable issues of plausibility, and the obviously religious foundations of the whole edifice, the Singularity was not a concept I could ever see myself getting behind. I will admit, that is, that I never succeeded in grasping the attraction of the thing, that I never quite came to understand how what it offered—the prospect of a bodiless existence as pure information, or run on some third-party human hardware—could ever be seen as salvation rather than perdition. If life had any meaning at all, my instinctive belief was that its meaning was animal, that it was inseparably bound up with birth, and reproduction, and death.

But more than any of this, the idea that technology would redeem us, that artificial intelligence would offer a solution to the suboptimal aspects of human existence, was incompatible with my basic outlook on life, with what little I happened to understand about the exceptionally destructive category of primates to which I belonged. Temperamentally and philosophically, I was and am a pessimist, and so it seemed to me that we were a great deal less likely to be redeemed than destroyed by the results of our own ingenuity. The planet, at this time, was on the verge of the sixth mass extinction since the first appearance of life on its surface, the first such extinction to be caused by the environmental impact of one of its native species.

Which is why when I began to read about the growing fear, in certain quarters, that a superhuman-level artificial intelligence might wipe humanity from the face of the earth, I felt that here, at least, was a vision of our technological future that appealed to my fatalistic disposition.

Such dire intimations were frequently to be encountered in the pages of broadsheet newspapers, as often as not illustrated by an apocalyptic image from the *Terminator* films—by a titanium-skulled killer robot staring down the reader with the glowing red points of its pitiless eyes. Elon Musk had spoken of AI as "our greatest existential threat," of its development as a technological means of "summoning the demon." ("Hope we're not just the biological boot loader for digital superintelligence," he'd tweeted in August of 2014. "Unfortunately, that is increasingly probable.") Peter Thiel had announced that "People are spending way too much time thinking about climate change, way too little thinking about AI." Stephen Hawking, meanwhile, had written an op-ed for *The Independent* in which he'd warned that success in this endeavor, while it would be "the biggest event in human history," might very well "also be the last, unless we learn to avoid the risks." Even Bill Gates had publicly admitted his disquiet, speaking of his inability to "understand why some people are not concerned."

Was I myself concerned? Yes and no. For all that they appealed to my inner core of pessimism, I was not much convinced by these end-time auguries, in large part because they seemed to me to be the obverse of Singulatarian prophecies about AI ushering in a new dispensation, in which human beings would ascend to unimaginable summits of knowledge and power, living eternally in the undimming light of Singularity's dawn. But I understood that my skepticism was more temperamental than logical, and that I knew almost nothing about the possibly excellent reasons for these fears, or the speculative technology that provoked them. And even if I couldn't quite bring myself to believe it, I was helplessly, morbidly fascinated by the idea that we might be on the verge of creating a machine that could wipe out our entire species, and by the notion that capitalism's great philosopher-kings–Musk, Thiel, Gates–were so publicly exercised about the Promethean dangers of that ideology's most cherished ideal. These dire warnings about AI were coming from what seemed the most unlikely of sources: not from Luddites or religious catastrophists, that is, but from the very people who seemed to most neatly personify our culture's reverence for machines.

One of the more remarkable phenomena in this area was the existence of a number of research institutes and think tanks substantially devoted to raising awareness about what was known as "existential risk"–the risk of absolute annihilation of the species, as distinct from mere catastrophes like climate change or nuclear war or global pandemics–and to running the algorithms on how we might avoid this particular fate. There was the Future of Humanity Institute in Oxford, and the Centre for Study of Existential Risk at the University of Cambridge, and the Machine Intelligence Research Institute in Berkeley, and the Future of Life Institute in Boston, which latter outfit featured on its board of scientific advisors not just prominent figures from science and technology like Musk and Hawking and the pioneering geneticist George Church, but also, for some reason, the beloved film actors Alan Alda and Morgan Freeman.

What was it that these people were referring to when they spoke of existential risk? What was the nature of the threat, the likelihood of its coming to pass? Were we talking about a *2001: A Space Odyssey* scenario, where a sentient computer undergoes some malfunction or other and does what it deems necessary to prevent anyone from shutting it down? Were we talking about a *Terminator* scenario, where a Skynettian matrix of superintelligent machines gains consciousness and either destroys or enslaves humanity in order to further its own particular goals? Certainly, if you were to take at face value the articles popping up about the looming threat of intelligent machines, and the dramatic utterances of savants like Thiel and Hawking, this would have been the sort of thing you'd have had in mind. They may not have been experts in AI, as such, but they were extremely clever men who knew a lot about science. And if these people were worried—along with Hawkeye from *M*A*S*H* and the guy who had played, among other personifications of serene wisdom, a scientist trying to prevent the Singularity in the 2014 film *Transcendence*—shouldn't we all be worrying with them?

One figure who loomed especially large over this whole area, its eschatologist-in-chief, was Nick Bostrom, the Swedish philosopher who, before he became known as the world's foremost prophet of technological disaster, had been one of the most prominent figures in the transhumanist movement, a cofounder of the World Transhumanist Association. In late 2014, Bostrom, director of the Future of Humanity Institute, had published a book called *Superintelligence: Paths, Dangers, Strategies,* in which he outlined the nature of the AI peril. For an academic text that made no significant gesture toward the casual reader, the book had been selling in unexpectedly high quantities, even to the point of appearing on the *New York Times* bestseller list. (The surge in sales was partly due to Elon Musk sternly advising his Twitter followers to read it.)

Even the most benign form of AI imaginable, the book suggested, could conceivably lead to the destruction of humanity. One of the more extreme hypothetical scenarios the book laid out, for instance, was one in which an AI is assigned the task of manufacturing paper clips in the most efficient and productive manner possible, at which point it sets about converting all the matter in the entire universe into paper clips and paper-clip-manufacturing facilities. The scenario was deliberately cartoonish, but as an example of the kind of ruthless logic we might be up against with an artificial superintelligence, its intent was entirely serious.

"I wouldn't describe myself these days as a transhumanist," Nick told me one evening over dinner at an Indian restaurant near the Future of Humanity Institute. Though he was married, his wife and young son were based in Canada, and he lived alone in Oxford. The arrangement involved frequent transatlantic flights, and regular Skype check-ins; regrettable though it was from a work-life balance point of view, it allowed him to focus on his research to a degree that would not otherwise have been possible. (He ate at this particular restaurant so frequently that the waiter had brought him a chicken curry without his having to make any explicit request for same.)

"I mean," he said, "I absolutely believe in the general principle of increasing human capacities. But I don't have much connection with the movement itself anymore. There is so much cheerleading of technology in transhumanism, so much unquestioning belief that things will just exponentially get better, and that the right attitude is just to let progress take its course. These are attitudes I have distanced myself from over the years."

Nick had become a kind of countertranshumanist of late; you couldn't reasonably accuse him of being a Luddite, but to the extent that he had made a name for himself, both within and outside academia, it was on the strength of his detailed forewarnings about where we might be bringing technology, and where it might take us.

"I still think," he said, "that within a few generations it will be possible to transform the substrate of our humanity. And I think artificial superintelligence will be the engine that drives that."

Like many transhumanists, Nick was fond of pointing out the vast disparity in processing power between human tissue and computer hardware. Neurons, for instance, fire at a rate of 200 hertz (or 200 times per second), whereas transistors operate at the level of gigahertz. Signals travel through our central nervous systems at a speed of about 100 meters per second, whereas computer signals can travel at the speed of light. The human brain is limited in size to the capacity of the human cranium, where it is technically possible to build computer processors the size of skyscrapers.

Such factors, he maintained, created the conditions for artificial superintelligence. And because of our tendency to conceive of intelligence within human parameters, we were likely to become complacent about the speed with which machine intelligence might exceed our own. Human-level AI might, that is, seem a very long way off for a very long time, and then be surpassed in an instant. In his book, Nick illustrates this point with a quotation from the AI safety theorist Eliezer Yudkowsky:

> AI might make an *apparently* sharp jump in intelligence purely as the result of anthropomorphism, the human tendency to think of "village idiot" and "Einstein" as the extreme ends of the intelligence scale, instead of nearly indistinguishable points on the scale of minds-in-general. Everything dumber than a dumb human may appear to us as simply "dumb." One imagines the "AI arrow" creeping steadily up the scale of intelligence, moving past mice and chimpanzees, with AI's still remaining "dumb" because AI's cannot speak fluent language or write science papers, and then the AI arrow crosses the tiny gap from infra-idiot to ultra-Einstein in the course of one month or some similarly short period.

At this point, the theory goes, things would change quite radically. And whether they would change for the better or for the worse is an open question. The fundamental risk, Nick argued, was not that superintelligent machines might be actively hostile toward their human creators, or antecedents, but that they would be indifferent. Humans, after all, weren't actively hostile toward most of the species we'd made extinct over the millennia of our ascendance; they simply weren't part of our design. The same could turn out to be true of superintelligent machines, which would stand in a similar kind of relationship to us as we ourselves did to the animals we bred for food, or the ones who fared little better for all that they had no direct dealings with us at all.

About the nature of the threat, he was keen to stress this point: that there would be no malice, no hatred, no vengeance on the part of the machines.

"I don't think," he said, "that I've ever seen a newspaper report on this topic that has not been illustrated by a publicity still from one of the *Terminator* films. The implication of this is always that robots will rebel against us because they resent our dominance, that they will rise up against us. This is not the case."

And this brought us back to the paper-clip scenario, the ridiculousness of which Nick freely acknowledged, but the point of which was that any harm we might come to from a superintelligent machine would not be the result of malevolence, or of any other humanlike motivation, but purely because our absence was an optimal condition in the pursuit of its particular goal.

"The AI does not hate you," as Yudkowsky had put it, "nor does it love you, but you are made out of atoms which it can use for something else."

One way of understanding this would be to listen to a recording of, say, Glenn Gould playing Bach's Goldberg Variations, and to try to experience the beauty of the music while also holding simultaneously in your mind a sense of the destruction that was wrought in the creation of the piano it is being played on: the trees felled, the elephants slaughtered, the human beings enslaved and killed in the interest of the ivory traders' profits. Neither the pianist nor the maker of the piano had any personal animosity toward the trees, the elephants, the enslaved men and women; but all were made of atoms that could be used for specific ends, for the making of money, the making of music. Which is to say that this machine which so terrifies a certain contingent of rationalists is perhaps not so different from us after all.

There is, among computer scientists working on AI, a particular reluctance to make predictions about how soon we might expect anything like a superhuman intelligence to emerge—even among those, by no means a majority, who believe such a prospect to be realistic. This has partly to do with a basic disinclination, among scientists in general, to make insufficiently substantiated claims that may wind up making them look foolish. But the reluctance also has a lot to do with the specific history of a field littered with conspicuous examples of people breezily underestimating the challenges involved. In the summer of 1956, before ideas about intelligent machines had begun to coalesce into anything like a discipline, a small group of scientists—leading figures in mathematics, cognitive science, electrical engineering, and computer science—gathered for a six-week-long workshop at Dartmouth College. The group included Marvin Minsky, Claude Shannon, and John McCarthy, men now seen as founders of AI. In a proposal to the Rockefeller Foundation, which funded the workshop, the group supplied the following rationale for its convening:

> We propose that a 2 month, 10 man study of artificial intelligence be carried out.... The study is to proceed on the basis of the conjecture that every aspect of learning or any other feature of intelligence can in principle be so precisely described that a machine can be made to simulate it. An attempt will be made to find how to make machines that use language, form abstractions and concepts, solve kinds of problems now reserved for humans, and improve themselves. We think that a significant advance can be made in one or more of these problems if a carefully selected group of scientists work on it together for a summer.

This sort of hubris has been an intermittent characteristic of AI research, and has led to a series of "AI winters"—periods of drastically decreased funding following bursts of intense enthusiasm about the imminent solution to some or other problem which then turned out to be much more complicated than imagined.

Repeated patterns, through the decades, of overpromising and underdelivering had led to a culture within AI whereby researchers were reluctant to look too far ahead. And this in turn has led to a difficulty in getting the field to engage seriously with the question of existential risk. Most developers working on AI did not want to be seen as making immoderate claims for the technology they were working on.

And whatever else you thought of it, this particular claim about the annihilation of humanity did lay itself open to the charge of immoderation.

Nate Soares raised a hand to his close-shaven head and tapped a finger smartly against the frontal plate of his monkish skull.

"Right now," he said, "the only way you can run a human being is on this quantity of meat."

We were talking, Nate Soares and I, about the benefits that might come with the advent of artificial superintelligence. For Nate, the most immediate benefit would be the ability to run a human being—to run, specifically, himself—on something other than this quantity of neural meat to which he was gesturing.

He was a sinewy, broad-shouldered man in his mid-twenties, with an air of tightly controlled calm; he wore a green T-shirt bearing the words "NATE THE GREAT," and as he sat back in his office chair and folded his legs at the knee, I noted that he was shoeless, and that his socks were mismatched, one plain blue, the other white and patterned with cogs and wheels.

The room was utterly featureless save for the chairs we were sitting on, and a whiteboard, and a desk, on which rested an open laptop and a single book, which I happened to note was a hardback copy of Bostrom's *Superintelligence.* This was Nate's office at the Machine Intelligence Research Institute, in Berkeley. The bareness of the space was a result, I assumed, of the fact that Nate had only just assumed his new role as executive director, having left a lucrative career as a software engineer at Google the previous year, and having subsequently risen swiftly up the ranks at MIRI. Nate's job had formerly been held by Eliezer Yudkowsky—the AI theorist Bostrom had cited on the quantum leap "from infra-idiot to ultra-Einstein"—who had founded MIRI in 2000. (It was originally called the Singularity Institute for Artificial Intelligence; the name was changed in 2013 to avoid confusion with

Singularity University, the Silicon Valley private college founded by Kurzweil and Peter Diamandis.)

I knew that Nate conceived of his task, and the task of MIRI, in starkly heroic terms, because I had read a number of articles he had written for the rationalist website Less Wrong in which he discussed his long-held desire to save the world from certain destruction. In one of these articles, I had read about his strict Catholic upbringing, his break with the faith in his teens, and his subsequent investment of his energies into "the passion, the fervor, the desire to optimize the future" through the power of rationality. Nate's rhetoric in these writings seemed to me a heightened performance of the Silicon Valley house style, in which every new social media platform or sharing economy start-up was announced with the fervent intention to "change the world."

At fourteen, he wrote, he became aware of the chaos of all human things, of a world around him "that couldn't coordinate," and he made a promise to himself. "I didn't promise to fix governments: that was a means to an end, a convenient solution for someone who didn't know how to look further out. I didn't promise to change the world, either: every little thing is a change, and not all changes are good. No, I promised to *save* the world. That promise still stands. The world sure as hell isn't going to save itself."

I was intrigued by the tone of Nate's writing, by the way in which it merged the language of logic with a kind of terse geek-romanticism: a strange, contradictory register which seemed to capture something essential about the idealization of pure reason that was such a prominent aspect of not just transhumanism, but of a broader culture of science and technology—something I had begun to think of as magical rationalism.

He spoke, now, of the great benefits that would come, all things being equal, with the advent of artificial superintelligence. By developing such a transformative technology, he said, we would essentially be delegating all future innovations—all scientific and technological progress—to the machine.

These claims were more or less standard among those in the tech world who believed that artificial superintelligence was a possibility. The problem-solving power of such a technology, properly harnessed, would lead to an enormous acceleration in the turnover of solutions and innovations, a state of permanent Copernican revolution. Questions that had troubled scientists for centuries would be solved in days, hours, minutes. Cures would be found for diseases that currently obliterated vast numbers of lives, while ingenious workarounds to overpopulation were simultaneously devised. To hear of such things is to imagine a God who had long since abdicated all obligations toward his creation making a triumphant return in the guise of a software, an alpha and omega of zeros and ones.

"With AI, what we're talking about is getting much quicker to the limit of what is physically possible," said Nate. "And one thing that will obviously become possible in this space is the uploading of human minds."

It was Nate's belief that, should we manage to evade annihilation by the machines, such a state of digital grace would inevitably be ours. There was, in his view, nothing mystical or fantastical about this picture of things, because there was, as he put it, "nothing special about carbon." Like everything else in nature—like trees, for instance, which he described as "nanotech machines that turn dirt and sunlight into more trees"—we were ourselves mechanisms.

Once we had enough computing power, he said, we would be able to simulate in full, down to the quantum level, everything that our brains were doing in their current form, the meat form.

Some version of this functionalist view of cognition was common among AI researchers, and was, in a sense, central to the entire project: the mind was a program distinguished not by its running on the sophisticated computing device of the brain, but by the operations it was capable of carrying out. (The project that people like Randal Koene and Todd Huffman were working toward now, for all its vast complexity and remoteness, could be achieved in the course of a long weekend by the kind of artificial superintelligence Nate was talking about.)

It was easy to forget, Nate continued, that as we sat here talking we were in fact using nanotech protein computers to conduct this exchange of ours, this transfer of data; and as he said this I wondered whether this conviction came so naturally to Nate, and to people like him, because their brains, their minds, functioned in the sort of logical, rigorously methodical way that made this metaphor seem intuitively accurate, made it seem, finally, not to be a metaphor at all. Whereas I myself had trouble thinking of my brain as a computer or any other kind of mechanism; if it were one, I'd be looking to replace it with a better model, because it was a profoundly inefficient device, prone to frequent crashes and dire miscalculations and lengthy meanderings on its way toward goals that it was, in the end, as likely as not to abandon anyway. Perhaps I was so resistant to this brain-as-computer idea because to accept it would be to necessarily adopt a model in which my own way of thinking was essentially a malfunction, a redundancy, a system failure.

There was something insidious about this tendency—of transhumanists, of Singularitarians, of techno-rationalists in general—to refer to human beings as though they were merely computers built from protein, to insist that the brain, as Minsky had put it, "happens to be a meat machine." (Something I'd read earlier that day on Nate's Twitter timeline, on which it was his custom to quote things overheard around the MIRI offices: "This is what happens when you run programs that fucked themselves into existence on computers made of meat.") This rhetoric was viscerally unpleasant, I felt, because it reduced the complexity and strangeness of human experience to a simplistic instrumentalist model of stimulus and response, and thereby opened an imaginative space—an *ideological* space, really—wherein humans could very well be replaced, versioned out by more powerful machines, because the fate of all technologies was, in the end, to be succeeded by some device that was more sophisticated, more useful, more effective in its execution of its given tasks. The whole point of technology per se was to make individual technologies redundant as quickly as possible. And in this techno-Darwinist view of our future, as much as we would be engineering our own evolution, we would be creating our own obsolescence. ("We are ourselves creating our own successors," wrote the English novelist Samuel Butler in 1863, in the wake of the Industrial Revolution, and four years after *On the Origin of Species*. "Man will become to the machine what the horse and dog are to man.")

But there was something else, too, something seemingly trivial but somehow more deeply disturbing: the disgust that arose from the conjunction of two ostensibly irreconcilable systems of imagery, that of flesh and that of machinery. And the reason, perhaps, that this union excited in me such an elemental disgust was that, like all taboos, it brought forth something that was unspeakable precisely because of its proximity to the truth: the truth, in this case, that we were in fact meat, and that the meat that we were was no more or less than the material of the machines that we were, equally and oppositely. And in this sense there really was nothing special about carbon, in the same way that there was nothing special, nothing *necessary*, about the plastic and glass and silicon of the iPhone on which I was recording Nate Soares saying that there was nothing special about carbon.

And so the best-case scenario of the Singularitarians, the version of the future in which we merge with artificial superintelligence and become immortal machines, was, for me, no more appealing, in fact maybe even *less* appealing, than the worst-case scenario, in which artificial superintelligence destroyed us all. And it was this latter scenario, the failure mode as opposed to the God mode, that I had come to learn about, and that I was supposed to be getting terrified about—as I felt confident that I would, in due course, once I was able to move on from feeling terrified about the best-case scenario.

As he spoke, Nate methodically uncapped and recapped with his thumb the lid of the red marker he was using to illustrate on the whiteboard certain facts and theories he was explaining to me: about how, for instance, once we got to the point of developing a human-level AI, it would then logically follow that AI would very soon be in a position to program further iterations of itself, igniting an exponentially blazing inferno of intelligence that would consume all of creation.

"Once we can automate computer science research and AI research," he said, "the feedback loop closes and you start having systems that can themselves build better systems."

This was perhaps the foundational article of faith in the AI community, the idea that lay beneath both the ecstasies of the Singularity and the terror of catastrophic existential risk. It was known as the intelligence explosion, a notion that had first been introduced by the British statistician I. J. Good, a former Bletchley Park cryptographer who went on to advise Stanley Kubrick on his vision of AI in *2001: A Space Odyssey*. In a paper called "Speculations Concerning the First Ultraintelligent Machine," delivered at a NASA conference in 1965, Good outlined the prospect of a strange and unsettling transformation that was likely to come with the advent of the first human-level AI. "Let an ultraintelligent machine be defined," he wrote, "as a machine that can far surpass all the intellectual activities of any man however clever. Since the design of machines is one of these intellectual activities, an ultraintelligent machine could design even better machines; there would then unquestionably be an 'intelligence explosion,' and the intelligence of man would be left far behind."

The idea, then, is that this thing of our creation would be the ultimate tool, the teleological end point of a trajectory that began with the hurling of the first spear—"the last invention that man need ever make." Good believed that such an invention would be necessary for the continued survival of the species, but that disaster could only be avoided if the machine were "docile enough to tell us how to keep it under control."

This mechanism, docile or otherwise, would be operating at an intellectual level so far above that of its human progenitors that its machinations, its mysterious ways, would be impossible for us to comprehend, in much the same way that our actions are, presumably, incomprehensible to the minds of the rats and monkeys we use in scientific experiments. And so this intelligence explosion would, in one way or another, be the end of the era of human dominance—and very possibly the end of human existence.

"It is unreasonable," as Minsky had put it, "to think that machines could become nearly as intelligent as we are and then stop, or to suppose that we will always be able to compete with them in wit or wisdom. Whether or not we could retain some sort of control of the machines, assuming that we would want to, the nature of activities or aspirations would be changed utterly by the presence on earth of intellectually superior beings."

This is, in effect, the basic idea of the Singularity, and of its dark underside of catastrophic existential risk. The word "singularity" is primarily a term from physics, where it refers to the point at the exact center of a black hole, where the density of matter becomes infinite, and in which the laws of space-time begin to break down.

"It gets very hard to predict the future once you have smarter-than-human things around," said Nate. "In the same way that it gets very hard for a chimp to predict what is going to happen because there are smarter-than-chimp things around. That's what the Singularity is: it's the point past which you expect you can't see."

Despite this conviction that the post-AI future would be even more difficult to predict than the garden-variety future—which was itself notoriously difficult to say anything useful or accurate about—it was Nate's reading of the situation that, whatever wound up happening, it was highly unlikely not to involve humanity being clicked and dragged to the trash can of history.

What he and his colleagues—at MIRI, at the Future of Humanity Institute, at the Future of Life Institute—were working to prevent was the creation of an artificial superintelligence that viewed us, its creators, as raw material that could be reconfigured into some more useful form (not necessarily paper clips). And the way Nate spoke about it, it was clear that he believed the odds to be stacked formidably high against success.

"To be clear," said Nate, "I do think that this is the shit that's gonna kill me."

I was, for some reason, startled to hear it put so bluntly. It made sense, obviously, that Nate would take this threat as seriously as he did. I knew well enough that this was not an intellectual game for people like him, that they truly believed this to be a very real possibility for the future. And yet the idea that he imagined himself more likely to be killed by an ingenious computer program than to die of cancer or heart disease or old age seemed, in the final analysis, basically insane. He had presumably arrived at this position by the most rational of routes–though I understood almost nothing of the mathematical symbols and logic trees he'd scrawled on the whiteboard for my benefit, I took them as evidence of such–and yet it seemed to me about as irrational a position as it was possible to occupy. Not for the first time, I was struck by the way in which absolute reason could serve as the faithful handmaiden of absolute lunacy. But then again, perhaps I was the one who was mad, or at least too dim-witted, too hopelessly uninformed to see the logic of this looming apocalypse.

"Do you really believe that?" I asked him. "Do you really believe that an AI is going to kill you?"

Nate nodded curtly, and clicked the cap back on his red marker.

"All of us," he said. "That's why I left Google. It's the most important thing in the world, by some distance. And unlike other catastrophic risks–like say climate change–it's dramatically underserved. There are thousands of person-years and billions of dollars being poured into the project of developing AI. And there are fewer than ten people in the world right now working full-time on safety. Four of whom are in this building. It's like there are thousands of people racing to be the first to develop nuclear fusion, and basically no one is working on containment. And we have got to get containment working. Because there are a lot of very clever people building something that, the way they are approaching it at present, will kill us all if they succeed."

"So as things stand," I said, "we're more likely to be wiped out by this technology than not, is what you're saying?"

"In the default scenario, yes," said Nate, and placed the red marker on his desk, balancing it upright like a prelaunch missile. His manner was, I felt, remarkably detached for someone who was talking about my death, or the death of my son, or of my future grandchildren—not to mention every other human being unlucky enough to be around for this imminent apocalypse. He seemed as though he were talking about some merely technical problem, some demanding, exacting bureaucratic challenge—as, in some sense, he was.

"I'm somewhat optimistic," he said, leaning back in his chair, "that if we raise more awareness about the problems, then with a couple more rapid steps in the direction of artificial intelligence, people will become much more worried that this stuff is close, and AI as a field will wake up to this. But without people like us pushing this agenda, the default path is surely doom."

For reasons I find difficult to identify, this term *default path* stayed with me all that morning, echoing quietly in my head as I left MIRI's offices and made for the BART station, and then as I hurtled westward through the darkness beneath the bay. I had not encountered the phrase before, but understood intuitively that it was a programming term of art transposed onto the larger text of the future. And this term *default path*—which, I later learned, referred to the list of directories in which an operating system seeks executable files according to a given command—seemed in this way to represent in miniature an entire view of reality: an assurance, reinforced by abstractions and repeated proofs, that the world operated as an arcane system of commands and actions, and that its destruction or salvation would be a consequence of rigorously pursued logic. It was exactly the sort of apocalypse, in other words, and exactly the sort of redemption, that a computer programmer would imagine.

—

And what was the nature of this rigorously pursued logic? What was it that was needed to prevent this apocalypse?

What was needed, first of all, was what was always needed: money, and clever people. And luckily, there were a number of people with sufficient money to fund a number of people who were sufficiently clever. A lot of MIRI's funding came in the form of smallish donations from concerned citizens—people working in tech, largely: programmers and software engineers and so on—but they also received generous endowments from billionaires like Peter Thiel and Elon Musk.

The week that I visited MIRI happened to coincide with a huge conference, held at Google's headquarters in Mountain View, organized by a group called Effective Altruism—a growing social movement, increasingly influential among Silicon Valley entrepreneurs and within the rationalist community, which characterized itself as "an intellectual movement that uses reason and evidence to improve the world as much as possible." (An effectively altruistic act, as opposed to an emotionally altruistic one, might involve a college student deciding that, rather than becoming a doctor and spending her career curing blindness in the developing world, her time would be better spent becoming a Wall Street hedge fund manager and donating enough of her income to charity to pay for several doctors to cure a great many more people of blindness.) The conference had substantially focused on questions of AI and existential risk. Thiel and Musk, who'd spoken on a panel at the conference along with Nick Bostrom, had been influenced by the moral metrics of Effective Altruism to donate large amounts of money to organizations focused on AI safety.

Effective Altruism had significant crossover, in terms of constituency, with the AI existential risk movement. (In fact, the Centre for Effective Altruism, the main international promoter of the movement, happened to occupy an office in Oxford just down the hall from the Future of Humanity Institute.)

It seemed to me odd, though not especially surprising, that a hypothetical danger arising from a still nonexistent technology would, for these billionaire entrepreneurs, be more worthy of investment than, say, clean water in the developing world or the problem of grotesque income inequality in their own country. It was, I learned, a question of return on investment—of time, and money, and effort. The person I learned this from was Viktoriya Krakovna, the Harvard mathematics PhD student who had cofounded—along with the MIT cosmologist Max Tegmark and Skype founder Jann Tallinn—the Future of Life Institute, which earlier that year had received an endowment of $10 million from Musk in order to establish a global research initiative aimed at averting AI catastrophe.

"It is about how much *bang* you get for your *buck*," she said, the American idiom rendered strange by her Ukrainian accent, with its percussive plosives, its throttled vowels. She and I and her husband, Janos, a Hungarian-Canadian mathematician and former research fellow at MIRI, were the only diners in an Indian restaurant on Berkeley's Shattuck Avenue, the kind of cavernously un-fancy setup that presumably tended to cater to drunken undergraduates. Viktoriya spoke between forkfuls of an extremely spicy chicken dish, which she consumed with impressive speed and efficiency. Her manner was confident but slightly remote, and, as with Nate, characterized by a minimal quantity of eye contact.

She and Janos were in the Bay Area for the Effective Altruism conference; they lived in Boston, in a kind of rationalist commune called Citadel; they had met ten years ago at a high school math camp, and had been together since.

"The concerns of existential risk fit into that value metric," elaborated Viktoriya. "If you consider the welfare of all future people, reducing the probability of a major future catastrophe can be a very high impact decision."

The Future of Life Institute was less focused than MIRI on the mathematical arcana of how a "friendly AI" might be engineered. The group, she said, functioned as "the outreach arm of this cluster of organizations," raising awareness about the seriousness of this problem. It was not the attention of the media or the general public for which FLI was campaigning, Viktoriya said, but rather that of AI researchers themselves, a constituency in which the idea of existential risk was only just beginning to be taken seriously.

One of the people who had been most instrumental in its being taken seriously was Stuart Russell, a professor of computer science at U.C. Berkeley who had, more or less literally, written the book on artificial intelligence. (He was the coauthor, with Google's research director Peter Norvig, of *Artificial Intelligence: A Modern Approach,* the book most widely used as a core AI text in university computer science courses.)

In 2014, Russell and three other scientists–Stephen Hawking, Max Tegmark, and Nobel laureate physicist Frank Wilczek–had published a stern warning, in of all venues *The Huffington Post,* about the dangers of AI. The idea, common among those working on AI, that because an artificial general intelligence is widely agreed to be several decades from realization we can just keep working on it and solve safety problems if and when they arise is one that Russell and his esteemed coauthors attack as fundamentally wrongheaded. "If a superior alien civilization sent us a text message saying 'We'll arrive in a few decades,' would we just reply, 'OK, call us when you get here–we'll leave the lights on'? Probably not–but this is more or less what is happening with AI."

The day after I had dinner with Viktoriya, I met Stuart at his office in Berkeley. Pretty much the first thing he did upon sitting me down was to open his laptop and turn it around toward me–a courtly gesture, oddly reminiscent of the serving of tea–so that I could read a few paragraphs of a paper called "Some Moral and Technical Consequences of Automation," by the cybernetics founder Norbert Wiener. The paper, originally published in the journal *Science* in 1960, was a brief exploration of the tendency of machines to develop, as they begin to learn, "unforeseen strategies at rates that baffle their programmers."

Stuart, an Englishman who radiated an aura of genial academic irony, directed me toward the last page of the paper, and sat in contemplative silence as I read on his screen the following passage: "If we use, to achieve our purposes, a mechanical agency with whose operation we cannot efficiently interfere once we have started it because the action is so fast and irrevocable that we have not the data to intervene before the action is complete, then we had better be quite sure that the purpose put into the machine is the purpose which we really desire and not merely a colorful imitation of it."

As I swiveled Stuart's laptop back in his direction, he said that the passage I had just read was as clear a statement as he'd encountered of the problem with AI, and of how that problem needed to be addressed. What we needed to be able to do, he said, was define exactly and unambiguously what it was we wanted from this technology. It was as straightforward as that, and as diabolically complex. It was not, he insisted, a question of machines going rogue, formulating their own goals and pursuing them at the expense of humanity, but rather a question of our own failure to communicate with sufficient clarity.

"I get a lot of mileage," he said, "out of the King Midas myth."

What King Midas wanted, presumably, was the selective ability to turn things into gold by touching them, but what he asked for (and what Dionysus famously granted him) was the inability *to avoid* turning things into gold by touching them. You could argue that his root problem was greed, but the proximate cause of his grief—which included, let's remember, the unwanted alchemical transmutations of not just all foodstuffs and beverages, but ultimately his own daughter—was that he was insufficiently clear in communicating his wishes.

The fundamental risk with AI, in Stuart's view, was no more or less than the fundamental difficulty in explicitly defining our own desires in a logically rigorous manner.

Imagine you have a massively powerful artificial intelligence, capable of solving the most vast and intractable scientific problems. Imagine you get in a room with this thing, and you tell it to eliminate cancer for once and for all. The computer will go about its work, and will quickly conclude that the most effective way to do so is to obliterate all species in which uncontrolled division of abnormal cells might potentially occur. Before you have a chance to realize your error, you've wiped out every sentient life form on earth, except for the artificial intelligence itself, which will have no reason not to believe it has successfully completed its task.

The AI researcher Stephen Omohundro, who sat with Stuart on MIRI's board of research advisors, published a 2008 paper outlining the dangers of goal-directed AI systems. The paper, entitled "The Basic AI Drives," contends that an AI trained upon even the most trivial of goals would, in the absence of extremely rigorous and complicated precautionary measures, present a very serious security risk. "Surely no harm could come from building a chess-playing robot, could it?" he asks, before briskly assuring us that a great deal of harm could in fact come from exactly that. "Without special precautions," he writes, "it will resist being turned off, will try to break into other machines and make copies of itself, and will try to acquire resources without regard for anyone else's safety. These potentially harmful behaviors will occur not because they were programmed in at the start, but because of the intrinsic nature of goal driven systems."

Because a chess-playing AI would be driven entirely by the maximization of its utility function (playing and winning chess), any scenario in which it might get turned off is one that it would be motivated to avoid, given that getting turned off would cause a drastic reduction in that utility function. "When a chess playing robot is destroyed," writes Omohundro, "it never plays chess again. Such outcomes will have very low utility and systems are likely to do just about anything to prevent them. So you build a chess playing robot thinking you can just turn it off should something go wrong. But, to your surprise, you find that it strenuously resists your attempts to turn it off."

So the challenge for the developers of artificial intelligence was, in this view, to design the technology so that it wouldn't mind getting turned off, and would otherwise behave in ways we found desirable. And the problem is that defining the sort of behavior we find desirable is not a straightforward matter. The phrase "human values" gets used a great deal in discussions of AI and existential risk, but its invocation is often qualified by an acknowledgment of the impossibility of any meaningfully accurate statement of said values. You might imagine, for instance, that you value the safety of your family above pretty much any other concern. And so you might think it sensible to instill, in a robot charged with the care of your children, the imperative that, whatever else it did or did not do, it must never cause those children to be put at risk of harm. This, in fact, is basically the first of Isaac Asimov's famous Three Laws of Robotics, which states: "A robot may not injure a human being or, through inaction, allow a human being to come to harm."

But the reality is that we're not quite as monomaniacally invested in the prevention of harm to our children as we imagine ourselves to be. A self-driving car that followed this instruction with absolute rigor would, for instance—given the nontrivial risk of getting into an accident on the way—decline to take your kids to the movies to see the latest computer-animated film about a young boy and his adventures with his robot pal.

One potential approach, most prominently proposed by Stuart himself, was that rather than attempting to write these implicit values and trade-offs into an AI's source code, the AI be programmed so that it learned by observing human behavior. "This is how we ourselves learn our value systems," he said. "Partly it's biological, in that, say, we don't like pain. Partly it's explicit, in that people tell you you shouldn't steal. But most of it is observing the behavior of other people, and inferring the values that are reflected in that. This is what machines need to be made to do."

When I asked him how far he felt we might be from a human-level artificial intelligence, Stuart was, in the customary manner of his profession, reluctant to offer predictions. The last time he'd made the mistake of alluding publicly to any sort of timeline had been the previous January at the World Economic Forum at Davos, where he sits on something called the Global Agenda Council on Artificial Intelligence and Robotics, and where he'd made a remark about AI exceeding human intelligence within the lifetime of his own children—the upshot of which, he said, had been a headline in the *Daily Telegraph* declaring that "'Sociopathic' Robots Could Overrun the Human Race Within a Generation."

This sort of phrasing suggested, certainly, a hysteria that was absent from Stuart's personal style. But in speaking with people involved in the AI safety campaign, I became aware of an internal contradiction: their complaints about the media's sensationalistic reporting of their claims were undermined by the fact that the claims themselves were already, sober language notwithstanding, about as sensational as it was possible for any claim to be. It was difficult to overplay something as inherently dramatic as the potential destruction of the entire human race, which is of course the main reason why the media—a category from which I did not presume to exclude myself—was so drawn to this whole business in the first place.

What Stuart was willing to say, however, was that human-level AI had come, in recent years, to seem "more imminent than it used to." Developments in machine learning like those spearheaded by DeepMind, the London-based AI start-up acquired by Google in 2014, seemed to him to mark an acceleration in the advancement toward something transformative. (Not long before I met Stuart, DeepMind had released a video demonstrating the result of an experiment in which an artificial neural network was set the task of maximizing its score in the classic Atari arcade game Breakout, in which the player controls a paddle at the bottom of a screen, with which they must break through a wall by bouncing a ball off it and thereby breaking its bricks. The video demonstrated the impressive speed and ingenuity with which the network had taught itself to play the game, developing new tactics in order to more effectively rack up points, and quickly surpassing all previous score records set by human beings.)

A computer batting its way to glory at a primitive arcade game was a long way from HAL 9000. What such neural networks had so far failed to master was the process of hierarchical decision making, which would necessitate looking ahead more than a few steps in the pursuit of a given task.

"Think about the kinds of decisions and actions that led you to sitting here in my office today," said Stuart, speaking so softly that I had to scoot my chair toward his desk and incline myself in his direction. "At the level of elementary moves, by which I mean actuations of your muscles and fingers and your tongue, your getting from Dublin to Berkeley might have involved something like five billion actions. But the really significant thing that humans are able to do in order to be competent in the real world–as opposed to, say, the world of a computer game or a chess program–is the ability to think in terms of higher-level actions. So rather than figuring out whether you should move this finger or that finger in this direction, or over that distance, you're instead figuring out whether you should fly United or British Airways to San Francisco, and whether you should then get an Uber or take the BART across the bay to Berkeley. You're able to think in these very large-scale chunks, and in that way you're able to construct futures that span billions of physical actions, most of which are entirely unconscious. That hierarchical decision making is a key component of human intelligence, and it's one we have yet to figure out how to implement in computers. But it's by no means unachievable, and once we do, we'll have made another major advance toward human-level AI."

After I returned from Berkeley, it seemed that every week or so the progress of artificial intelligence had passed some new milestone. I would open up Twitter or Facebook, and my timelines–flows of information that were themselves controlled by the tidal force of hidden algorithms–would contain a strange and unsettling story about the ceding of some or other human territory to machine intelligence. I read that a musical was about to open in London's West End, with a story and music and words all written entirely by an AI software called Android Lloyd Webber. I read that an AI called AlphaGo–also the work of Google's DeepMind–had beaten a human grandmaster of Go, an ancient Chinese strategy board game that was exponentially more complex, in terms of possible moves, than chess. I read that a book written by a computer program had made it through the first stage of a Japanese literary award open to works written by both humans and AIs, and I thought of the professional futurist I had talked to in the pub in Bloomsbury after Anders's talk, and his suggestion that works of literature would come increasingly to be written by machines.

I was unsure how to feel about all of this. In one sense, I was less disturbed by the question of what the existence of computer-generated novels or musicals might mean for the future of humanity than by the thought of having to read such a book, or endure such a performance. And neither had I taken any special pride in the primacy of my species at strategy board games, and so I found it hard to get excited about the ascendancy of AlphaGo, which seemed to me like a case of computers merely getting better at what they'd always been good at anyway, which was the rapid and thorough calculation of logical outcomes–a highly sophisticated search algorithm. But in another sense, it seemed reasonable to assume that these AIs would only get better at doing what they already did: that the West End musicals and sci-fi books would become incrementally less shit over time, and that more and more complicated tasks would be performed more and more efficiently by machines.

At times, it seemed to me perfectly obvious that the whole existential risk idea was a narcissistic fantasy of heroism and control–

a grandiose delusion, on the part of computer programmers and tech entrepreneurs and other cloistered egomaniacal geeks, that the fate of the species lay in their hands: a ludicrous binary eschatology whereby we would be either destroyed by bad code or saved by good code. The whole thing seemed, at such moments, so childish as to barely be worth thinking about, except as an object lesson in the idiocy of a particular kind of cleverness.

But there were other occasions when I would become convinced that *I* was the one who was deluded, and that Nate Soares, for instance, was absolutely, terrifyingly right: that thousands of the world's smartest people were spending their days using the world's most sophisticated technology to build something that would destroy us all. It seemed, if not quite plausible, on some level intuitively, poetically, mythologically *right*. This was what we did as a species, after all: we built ingenious devices, and we destroyed things.

A Short Note on the First Robots

IN PRAGUE, ON the evening of January 25, 1921, human beings were first introduced to robots, and shortly thereafter to the elimination of their own species by same. This event occurred in the Czech National Theatre on the opening night of Karel Čapek's play *R.U.R.* The title stood for "Rossum's Universal Robots," and marked the first ever usage of a term—derived from the Czech word "robota," meaning "forced labor"—which would quickly become a convergence point in the intersecting mythologies of science fiction and capitalism. Visually, Čapek's robots have less in common with later canonical representations of gleaming metallic humanoids—the more or less direct lineage from Fritz Lang's *Metropolis* to George Lucas's *Star Wars* to James Cameron's *Terminator*—than with *Blade Runner*'s uncannily convincing replicants. They look more or less indistinguishable, that is, from humans; they are creatures not of circuitry and metal, but of flesh, or a fleshlike substance—produced in a series of "mixing vats," one for each organ and body part, from a mysterious compound referred to as "batter." The play is itself a strange, viscous concoction of sci-fi fable, political allegory, and social satire, whose polemical intentions rest uneasily between a critique of capitalist greed and an anticommunist fear of the organized mob.

Čapek's robots are "artificial people" created for the purpose of increased industrial productivity, and represent, through the prism of the profit motive, an oppressively reductive view of human meaning. In the play's first scene, a man named Domin, the manager of the robot production plant in which the action is set, is afforded a monologue, as flagrantly didactic as his own name, on how the inventor of these machines (the eponymous Rossum) created "a worker with the smallest number of needs, but to do so he had to simplify him. He chucked everything not directly related to work, and in so doing he pretty much discarded the human being and created the Robot." These robots are, like Frankenstein's monster before them, created fully formed, and are ready to begin work immediately. "The human machine," he explains, "was hopelessly imperfect.... Nature had no grasp of the modern rate of work. From a technical standpoint the whole of childhood is pure nonsense. Simply wasted time. An untenable waste of time."

The explicit ideology behind the creation of these robots seems now, in its contradictory mix of ruthless corporatism and messianic rhetoric, strangely suggestive of Silicon Valley techno-progressivism, and of some of the more extravagant predictions about AI. Domin—who the stage directions specify is seated at "a large American desk" backed by printed posters bearing messages like *"The Cheapest Labor: Rossum's Robots"*—insists that this technology will eradicate poverty entirely, that although people will be out of work, everything will be done by machines, and they will be free to live in pursuit of their own self-perfection. "Oh Adam, Adam!" he says, "no longer will you have to earn your bread by the sweat of your brow; you will return to Paradise, where you were nourished by the hand of God."

As is customary with such enterprises, it doesn't work out: the robots, having proliferated greatly by the play's second act and having in many cases received military training in technologically forward-thinking European states, decide they no longer consent to be ruled over by a species they view as inferior, and so resolve to eradicate that species, which task they set about with precisely the sort of efficiency and singularity of purpose so highly valued by their human creators.

Aside from its allegorical depiction of capitalism's mechanization of its subjects, the play, in its blunt fashion, animates an associated Promethean fear of technologies intended to replicate human life. The rising up of the robots, and the almost total elimination of humanity that follows, is presented as no more or less than divine vengeance—as the damnation that is the inevitable end of any attempt to recover Paradise.

Čapek's robots appear as distorted doubles of ourselves. They are, as specified in the stage directions, "dressed like people," their faces "expressionless" and their eyes "fixed"; they bring to mind in this way both automata and corpses, possessed of the familiar otherness of the living dead. The robot genocide of the play's third act, written in the immediate aftermath of the First World War, is more or less explicitly positioned as a reflection of Europe's imperial ceremony of technologically enhanced bloodshed, as an acting out of the "human values" the robots have learned from their creators. When Alquist, the last surviving human, asks the leader of the robots why they have wiped out every other living person, he is told, "You have to kill and rule if you want to be like people. Read history! Read people's books! You have to conquer and murder if you want to be people!"

The robots in Čapek's play are a nightmare of future technology born out of the terror of present humanity. "Nothing is stranger to man," as Alquist puts it, "than his own image." And so these first fictional robots reveal the way in which our technologies reflect back on us the values out of which they are forged—"more horrid even," as Frankenstein's monster says of himself, "from the very resemblance."

What is this fascination of ours, this obsession with our own destruction by consequence of our ingenuity? Daedalus, that old artificer, is the symbol and spirit of this understanding of ourselves, of our ambitions—the shadow of the waxwing slain, cast darkly across history, plummeting.

Artificial superintelligence, I was repeatedly told, was a dangerous prospect precisely because of how unlike us it would be, how inhuman, how immune to anger and hatred and empathy. But a sidelong reading of this cryptic eschatology suggests itself: perhaps this fear of what might be done to us by our most sophisticated technology, by our last invention, is a kind of sublimated horror at what we have already done to the world, to ourselves. We are already, many of us, controlled in ways we barely reflect on by machineries we barely understand; and the history of science and technology, at its best and at its worst, is a history of the conquest of nature, of the curing of diseases and the eradication of vast numbers of species. And so perhaps this apparition of a vengeance wreaked upon us by our own evolutionary successor is an expression of an existential shame. Perhaps it is a transfigured form of original sin, a return of the repressed, a neurotic avatar of some deeper terror. Nothing stranger to us than our own image.

Mere Machines

ROBOTS WERE, IN one way or another, our future. This much I had been given to understand by the transhumanists I had spoken to, the foretellings of the future I had heard. If you were to believe Randal Koene or Natasha Vita-More or Nate Soares, we were ourselves going to *be* robots, our minds uploaded to machines much stronger and more efficient than our primate bodies. Or we were going to live increasingly among machines, ceding more and more of our work and our lives to their authority and order. Or they were going to make us obsolete and replace us as a species.

I would sit at breakfast and watch my son play with the little windup robot toy I had brought him back from San Francisco, its Frankensteinian lurch across the table toward the fruit bowl, and I would wonder what role actual robots would play in his future—how many of the careers I imagined for him would even exist in twenty years' time, how many of them would by then have been lost to total automation, the final dream of corporate techno-capitalism.

He intercepted me in the hallway one day, having watched two or three back-to-back episodes of a cartoon called *Animal Mechanicals*.

"I'm a walking machine," he said, and shuffled robotically in a circle about my legs.

It seemed a strange thing to say, but then again more or less everything he said was a strange thing to say.

I'd been thinking a lot about robots, but I had never seen an actual robot. Not, as it were, in the flesh; not, as it were, in action. I'd been thinking a lot about robots, but I didn't know what it was, exactly, I'd been thinking about. And then I heard about a thing called the DARPA Robotics Challenge, an event at which the world's foremost robotics engineers gathered to set their creations against one another in a series of trials designed to test their performance in human environments, in situations of extreme peril and stress.

I'd seen it described by *The New York Times* as "Woodstock for robots," and I wanted to witness it for myself.

Aside from the considerable prestige of winning the DARPA Robotics Challenge, the ultimate victor, or its makers, stood to receive a million dollars in prize money from the competition's eponymous benefactor. DARPA, the Defense Advanced Research Projects Agency, is the wing of the Pentagon responsible for the development of emerging technology for military purposes. The organization, which was set up by President Dwight Eisenhower in 1958 as a response to the Soviet launch of Sputnik, has a history of originating transformative technologies. With its ARPANET project in the late 1960s, for instance, DARPA laid down the technical foundations of the Internet. GPS, the very technology by which my Uber driver efficiently piloted me from West Hollywood to Pomona for the event, was also a DARPA innovation: a tool of war by which I increasingly navigate the world. According to its own strategic plan, DARPA's explicit purpose is "to prevent technological surprise to the US, but also to create technological surprise for our enemies."

The more fascinated I became by the transhumanist movement, and the more I learned about the various innovations to which these people were tethering their hopes for a posthuman future, the more I kept coming across references to DARPA, and its funding of these potentially transformative technologies: brain-computer interfaces, cognitive prosthetics, augmented cognition, cortical modems, bioengineered bacteria, and so forth. DARPA's overall objective these days, it seemed, was transcending the limits of the human body; specifically the human bodies of American soldiers.

The event at the Fairplex, a vast outdoor events venue in Pomona, constituted the finals of a contest that had been running since 2012; but it was also a celebration of the fruitful union of defense and corporate and scientific interests, a military-industrial pageantry. The whole thing was aimed, in the words of the event's director, Gill Pratt, at expediting the development of semiautonomous robots capable of performing "complex tasks in dangerous, degraded, human-engineered environments."

Its immediate inspiration was the hydrogen explosion at the Fukushima nuclear power plant in 2011, a catastrophe that might have been significantly mitigated by robots capable of negotiating environments designed for human bodies. At a briefing that morning, a press guy named Bradley addressed an art deco ballroom full of DARPA employees and media people, informing them that humanitarian disaster relief was "one of the core missions of the US military," and that humanoid robots were an increasingly significant dimension of this mission.

"If you could double the number of joints in your arm," he said, "think of all the different ways you could open a door."

I was given a press pack that contained a color printout illustrating the eight tasks the robots were obliged to complete: Drive Utility Vehicle, Exit Utility Vehicle, Open and Go Through Door, Locate and Close Valve, Cut Through Wall, Surprise Task, Rubble (Clear Debris *or* Walk on Rough Terrain), and Climb Stairs.

What I saw, when I took my seat in the grandstand overlooking Fairplex's racetrack, was a series of stage sets representing several identical iterations of a kind of generic industrial disaster zone: brick walls, artificially distressed "DANGER: HIGH VOLTAGE" signs, cartoonishly large red levers (this, it turned out, was the day's Surprise Task), wall-mounted valve wheels, gauntlets of heaped and broken concrete. Each of these mock-ups was the setting in which a given robot had to accomplish a suite of tasks that, for any minimally competent human being, would be utterly straightforward, but which for these rude mechanicals constituted a rigorously demanding sequence of technical challenges.

A small red utility vehicle was being driven along a sandy track, haltingly navigated between a pair of red plastic safety barriers, by a robot with what seemed to be a camera rotating slowly where its face should be. The robot was not sitting in the car, but rather standing on a footplate by the passenger side door, reaching across the interior of the vehicle to steer with a long, clawlike arm. The smell of hot popcorn drifted upward from the concourse below, lingering in the warm Californian air like an atmospheric irony, and a Jumbotron directly in front of me displayed a blandly handsome announcer seated behind a curved desk emblazoned with DARPA's logo: a sports broadcast mise-en-scène from some speculative future, vaguely fascist, in which the machinery of national defense had become a spectacle of mass entertainment.

"He's really chugging along here," said the man. At the opposite end of the desk was a smiling woman with short silver hair, wearing a blue polo shirt, likewise branded; this was Arati Prabhakar, DARPA's director.

"Wow, that's terrific to see!" she said.

I found it difficult to accommodate the sight of this pleasant-looking woman, smiling fondly at the driving robot, with what I knew about the organization she led. When I thought of DARPA, I thought, among other things, of its administration of the so-called Information Awareness Office, exposed by the former CIA employee Edward Snowden as a mass surveillance operation organized around a database for the collection and storage of the personal information (emails, telephone records, social networking messages, credit card and banking transactions) of every single resident of the United States, along with those of many other countries, all of which was accomplished by tapping into the user data of marquee-name tech companies like

Facebook, Apple, Microsoft, Skype, Google—the corporate proprietors of the sum of things that might be factually and usefully said about you, your information.

"Look at him go!" said Prabhakar now, as the robot rounded the second safety barrier, bringing the car over a line in the sand, and gently to a halt in front of the door through which the industrial disaster zone stage set was to be accessed by means of knob-turning. "That's fun!"

"This is like the Super Bowl of robots," said the announcer. "It's tremendously exciting."

"Yes," said Prabhakar, chuckling. "We hope to spark fires. That's what we're doing here: trying to create this groundswell of excitement about robotics, so as to push the development of the technology forward."

The disaster-response application of this technology was the narrative that the DARPA brass kept pushing throughout the weekend, but Prabhakar was not reticent when it came to the topic of the eventual military deployment of these machines. "In the military context," she told an interviewer, "our warfighters have to do incredibly dangerous tasks as a core part of their missions. As robotics technology advances and we can harvest it to help alleviate those challenges for our warfighters, that's absolutely something we'll be looking to do."

The robot successfully dismounted the car, proceeding at a slight crouch, and with exaggerated caution, toward the door; these movements it performed in the manner of a prodigiously shitfaced man intent on demonstrating that he had only had a couple of sherries with dinner. And then for ten minutes, perhaps fifteen, precisely nothing happened. Perhaps there had been a break in the wireless communication channels linking the robot with its team of engineers, huddled around their bank of screens in a hangarlike building off to the rear of the racetrack; these network dropouts were a deliberate tactic on DARPA's part, a built-in component of the competition intended to test the robots' level of autonomy, their ability to go about their business without need of remote micromanagement.

The robot I was watching was, I learned from the announcer, the creation of Pensacola's Florida Institute for Human and Machine Cognition, and his—or its—name was Running Man. (This, I realized with a small numinous thrill, was the very robot that had appeared on the cover of that week's *Time*, a copy of which I'd bought in Heathrow Airport the previous morning, before boarding a plane whose in-flight entertainment package, I may as well tell you, included no fewer than four robo-centric film options: *Big Hero 6*, an animated children's movie about a young boy and his robot friend; *Ex Machina*, an enjoyably creepy film about a Dr. Moreauvian Silicon Valley billionaire who holes himself up in a remote and hyper-secure mansion with a coterie of beautiful female sex-bots; *Chappie*, a just-about-watchable South African sci-fi romp about a police robot that gains sentience and falls in with a gang of armed robbers; and a bargain-basement B movie called *Robot Overlords*, which concerned the invasion of the Earth by tyrannical robots from outer space, and which starred a scenery-devouring Sir Ben Kingsley, whose fee I would guess accounted for much of the film's budget.) Running Man continued not running, or walking, or moving in any perceptible way at all, for an impressively long time. And then he, or it, did conspicuously move: the arm that had been poised in front of the doorknob finally made contact with its target and caused it to turn, and all of a sudden the door had swung inward, and the robot was beginning to make its cautious piston-legged way into the room, and the assembled crowd of tech enthusiasts and DARPA employees and U.S. Marines and young fathers with their children burst into a volley of cheers and applause, and the announcer, in exactly the kind of decorously enthused tone you'd expect from a golf commentator on ESPN, declared "another point for Running Man, who is now making his way—very smoothly, I might add—through the room." On the Jumbotron behind the stage, the image of the moving robot gave way to a gigantic animated graphic confirming that Running Man, and its team of furiously industrious backstage engineers, had just successfully completed the door-opening-and-room-entering stage of the course, and had indeed accordingly been awarded a point.

In the row in front of me, a boy of about ten turned to his father and announced in a casually authoritative tone that this was "by far one of the most interesting robots I've ever seen."

All that warm and festive Friday morning, I watched robots of various designs and abilities attempt to complete these tasks, and was entertained beyond all reasonable expectation by the spectacle. This was due in part to the competitive structure and sporting tenor of the event: the scoreboard and the live commentary, the Jumbotron and sideline engineer interviews, the pervasive American balm of hot dogs and popcorn. But mostly, it was due to the unanticipated element of slapstick, the weird confluence of high tech and low comedy.

I saw a robot stand perfectly still for fifteen minutes before succumbing to a powerful tremor at the knees and pitching over sideways, as though felled by some terrible seizure in its circuitry; I saw a robot finally succeeding in opening a door, and then falling through the frame to land flat on its serene titanium face; I saw a robot reach out to turn a valve wheel, miss its target by two or three inches, and claw the empty air in a counterclockwise direction before plummeting headfirst along the vector of its turning. I saw a great many robots collapsing backward in the attempt to walk up a stairs to nowhere, and a great many more brought low by the gauntlet of rubble, to be stretchered out on literal stretchers by teams of hard-hatted engineers.

The obvious implication of all of this, and on some level the rationale for the whole event, was that, although our technologies tend to be quite good at performing tasks beyond our own capabilities—flying at great heights and speeds, for instance, or processing massive amounts of data—they tend to be quite bad at doing things we do without thinking, like walking and picking up objects and opening doors, things that are really extraordinarily complex and demanding.*

* This is known, apparently, as Moravec's Paradox, after robotics professor Hans Moravec's observation that "it is comparatively easy to make computers exhibit adult level performance on intelligence tests or playing checkers, and difficult or impossible to give them the skills of a one-year-old when it comes to perception and mobility."

As such, the driving component of the competition proved to be far less problematic than the subsequent exiting of the vehicle—a distinct task in its own right, ruefully referred to by the engineers as "egress." As the announcer suavely pointed out, egress was so difficult for the robots that many of the competing teams chose to go ahead and skip the whole thing, sacrificing a point to save time and money. (As amusing as it is to watch a robot pitch over headfirst while attempting to extricate itself from the driver's seat of a doorless SUV, these pratfalls can amount to a significant setback for their wranglers. The robots who made it as far as the finals cost anything from several hundred thousand to millions of dollars to make, and so reparation of the damage caused by falls was an expensive business, and also a time-consuming one.)

But as I watched these visions of technological ingenuity and military-industrial power falling arse over tip in the effort to perform apparently simple tasks, I began to wonder whether this incidental spectacle of physical comedy was not, in some sense, central to the whole enterprise—whether the unspoken or unconscious intention here might not be to raise the robots to the level of humans not merely in terms of physical capability, but to elevate them in other ways, too. Because there is something deeply human, and humane, about the relationship between the body undergoing a pratfall and the body observing. There is cruelty in this laughter, but also empathy. These robots are literally inhuman, and yet I react no differently to their stumblings and topplings than I would to the pratfalls of a fellow human. I don't imagine I would laugh at the spectacle of a toaster falling out of an SUV, or a semiautomatic rifle pitching over sideways from an upright position, but there is something about these machines, their human form, with which it is possible to identify sufficiently to make their falling deeply, horribly funny.

I thought of a line from Henri Bergson's book *Laughter: An Essay on the Meaning of the Comic*—a line that, perhaps precisely because I had never quite understood it, had stayed with me in the years since I had first read it. "The attitudes, gestures and movements of the human body," he wrote, "are laughable in exact proportion as that body reminds us of a mere machine." I found the robots' pratfalls comical, in other words, not simply because in their forms and their failures they resembled humans, but because they reflected the strange sense in which humans were themselves mere machines.

Not everyone found these robots falling on their faces so amusing. I saw one of the stewards, a woman in her early twenties with a blue DRC STAFF T-shirt, greet a colleague on the steps. "Did you see that robot who fell over just there? It was really sad to watch." Her colleague agreed; it made her sad, too. "I felt terrible for him."

On the Jumbotron, a robot accomplished a flawless exit from its SUV, and began its approach to the door.

"Impressive performance now from Momaro over on that yellow course," said the announcer. "Just a *very* impressive egress."

At midday, the robots broke for lunch. There was a burst of enthusiastic applause, and the robust drumbeat and rumbling bass line of the Foo Fighters' "My Hero" blasted from the PA system, as footage of some of the morning's more audacious feats of vehicle-exiting and door-opening and lever-pulling played on the Jumbotron.

I noticed, then, a dark form hovering above the racetrack, high and lonesome as a buzzard in the shimmering heat of noon. It was a small drone: another reminder of DARPA's history of innovation in unmanned combat, its wider project of panoptic surveillance, signature strikes, screamings across the sky. As I watched the silent machine rise against the backdrop of the San Jose Hills, flashing in the sun with a glitter of knives, I felt with sudden force the schizoid strangeness of the event I had come to witness, which seemed designed both to reassure the world of DARPA's humanitarian intentions and to expedite the development of technologies that would, in the fullness of time, be trained upon kill zones far from this fairground, with its onsite Sheraton, its conference center, its dedicated RV parking lot.

I looked around, and saw the crowd—the families with their young children; the clusters of twenty- and thirty-something programmers; the uniformed Marines who were themselves human components of the machinery of government referred to by Hobbes as the "great Leviathan called a Common-Wealth or State which is but an Artificiall Man"—filtering downward out of the grandstand toward the burger concessions and hot dog carts, and was overtaken by a sudden bleak intimation of technology as an instrument of human perversity, in the service of power and money and war.

Out in the fairground, in the Technology Exposition area where the industry had come to set out its various stalls, the general understanding was that robots were the future. The word for what was going on here would most likely be something like "outreach," or "engagement." Walking under a large canvas DARPA banner ("Thank You for Cheering Us On!"), I entered a kind of scaffolded tunnel that housed a "DARPA Through the Decades" exhibit. Highlighted here were some of the organization's major accomplishments, among the more recent of which were the 2003 launch of the X-45A, an early prototype of the Predator and Reaper drones responsible for the deaths of hundreds of Pakistani civilians and children, and a monstrous unmanned armored ground vehicle named, with admirable frankness, "The Crusher."

Further on, I passed a black quadruped robot in a glass display cabinet, a nightmarish pastiche of a Damien Hirst installation. The encased specimen was a creature known as Cheetah, developed with DARPA funding by Boston Dynamics, an industry-leading robotics laboratory that had been acquired by Google in 2013. This robot was capable of running at 28.3 miles per hour, faster than any recorded human. I had seen it in action on YouTube—itself a wholly owned subsidiary of Google—and it was somehow thrilling and abominable: this rough beast, its hour come at last, emerging at an uncanny gallop from some final merger of corporate and state power in the crucible of technology.

I walked on, and saw a tall, sickly looking young man wearing dark sunglasses, a black fedora, and a black suit with a vaguely clerical purple silk shirt. A toy monkey was perched on his shoulder, and in his black leather-gloved hands he held a small device with which he was controlling an arachnoid robot roughly the size of a bull terrier. Standing next to him was another man, who wore a laminated DARPA lanyard around his neck, and who was presumably the father of the sun-hatted toddler, some eight or ten feet away, being chased in a widening circle by the mechanical spider.

At the stall of a company called Softbank Robotics, a Frenchman was attempting to convince a four-foot humanoid to hug a three-year-old girl.

"Pepper," he said. "Please hug the little girl."

"I'm sorry," said Pepper, in an appealingly childlike voice lightly inflected with a Japanese accent and genuine regret. "I didn't understand."

"Pepper," said the Frenchman, with elaborate clarity and forbearance. "Can you please give this little girl a hug."

The little girl in question, who was sullen and silent and clutching the leg of her father, did not look much like she wanted Pepper to give her a hug.

"I'm sorry," said Pepper again. "I didn't understand."

I felt a sudden surge of compassion for this winsome creature, with its huge innocent eyes, its touchscreen chest, its beautiful human failure to understand.

The Frenchman smiled tightly, and bent down to the side of the robot's head, where its auditory receptors were located.

"Pepper! Please! Hug! The girl!"

Pepper at last raised her arms, and made her wheeled approach to the child, who then gave herself up, tentatively and with obvious misgivings, to the robot's embrace, before quickly backing out of the whole deal and returning to the shelter of her father's legs.

The Frenchman explained to me that Pepper was a customer service humanoid, designed to "interface with people in a natural and social manner." It was capable, apparently, of feeling emotions ranging from joy to sadness to anger to doubt, its "mood" influenced by data received through touch sensors and cameras.

"It is mostly for greeting people when they come in. At a mobile phone store, for instance, it will come to you and ask you if you need something, and maybe explain you some special offers that the mobile phone store is running. It will give you a fist-bump, or maybe a hug. As you can see, we are still perfecting this, but we are close. You would be surprised how difficult it is to solve the problem of hugging."

I asked him whether robots like these were intended eventually to replace the human beings who currently worked in mobile phone stores, and he told me that, although this was a likely eventuality of the progress of robotics, Pepper's immediate function was purely a "social and emotional" one: she was a kind of corporate ambassador from the future, intended to put customers at their ease in the presence of humanoid robots.

"We need to break that barrier first," he said. "People will eventually become comfortable."

I did not doubt that this was true. Already, we had become comfortable with automated checkouts at supermarkets, with touchscreens and instructions from computerized voices where previously there would have been a human being, earning a paycheck.

Earlier that week, in Seattle, Amazon had held a robotics competition of its own. The Amazon Picking Challenge set companies the task of developing a robot capable of replacing its human stock pickers. And you could see how this would make sense for Amazon, a company that had long been known for its poor treatment of its warehouse workers, and for its monomaniacal focus on the elimination of every kind of middleman—of booksellers, editors, publishers, postal workers, couriers. (Amazon was, at that point, on the verge of launching a drone delivery program, whereby consumer goods made and packaged by robots could be delivered into your hands by an unmanned carrier-drone within thirty minutes of you placing your order.) Robots don't need toilet breaks, and drones don't get tired, and neither are likely to form unions.

And so this seemed like the ultimate fulfillment of the logic of techno-capitalism: the outright ownership not just of the means of production, but of the labor force itself. Čapek's term "robot" was, after all, taken from the Czech word for "forced labor." The image and valence of the human body has always shaped how we think about machines; humans have always succeeded in reducing the bodies of other humans to mechanisms, components in systems of their own design. As Lewis Mumford put it in his book *Technics and Civilization*, written during the early years of the Great Depression:

> Long before the peoples of the Western World turned to the machine, mechanism as an element in social life had come into existence. Before inventors created engines to take the place of men, the leaders of men had drilled and regimented multitudes of human beings: they had discovered how to reduce men to machines. The slaves and peasants who hauled the stones for the pyramids, pulling in rhythm to the crack of the whip, the slaves working in the Roman galley, each man chained to his seat and unable to perform any other motion than the limited mechanical one, the order and march and system of attack of the Macedonian phalanx–these were all machine phenomena. Whatever limits the actions of human beings to their bare mechanical elements belongs to the physiology, if not the mechanics, of the machine age.

Recently, on the website of the World Economic Forum, I had seen a list of the "20 Jobs That Robots Are Most Likely to Take Over." Jobs with a 95 percent or higher chance of their practitioners being made obsolete by machines within twenty years included postal workers, jewelers, chefs, corporate bookkeepers, legal secretaries, credit analysts, loan officers, bank tellers, tax accountants, and drivers.

This last occupation, which was the largest category of employment for American men, was particularly ripe for automated disruption. The original DARPA Grand Challenge, held in 2004 to stimulate the development of driverless vehicles, was a 150-mile race across the Mojave Desert from Barstow to the Nevada border. The event was a fiasco: not one of the robotic vehicles even came close to finishing the route. The car that got the farthest from the starting pistol made it just under seven and a half miles before finally coming to grief on a large rock, and DARPA declined to award its million-dollar prize.

But when the race was held again the following year, five cars finished the route, and the winning team went on to form the nucleus of Google's Self-Driving Car Project, under the auspices of which, even now, California's roads were being successfully navigated by vehicles unguided by human hands, luxury ghostmobiles on the decaying highways, an advance guard of an automated future. Uber, the drive-sharing service that had seriously damaged the taxi sector in recent years, was already speaking openly about its plans to replace all of its drivers with automated cars as soon as the technology allowed. At a conference in 2014, the company's preeminently obnoxious CEO Travis Kalanick had explained that "the reason Uber could be expensive is because you're not just paying for the car, you're paying for the other dude in the car. When there's no other dude in the car, the cost of taking an Uber anywhere becomes cheaper than owning a vehicle." When asked how he might explain to these other dudes the reality of their obsolescence, their versioning out, he said this: "Look, this is the way of the world, and the world isn't always great. We all have to find ways to change the world." Kalanick, I had heard, was here in Pomona today, in search of further ways to change a world that was increasingly his to change.

The Frenchman asked if I would like a hug from Pepper, and I assented as much out of politeness as journalistic rigor.

"Pepper," he said. "This man would like a hug."

I fancied that I detected something like ambivalence in Pepper's impassive gaze; but she raised her arms and I bent toward her, and suffered her to enfold me in her unnatural clasp. It was, frankly, an underwhelming experience; I felt that we were both, in our own ways, phoning it in. I patted her on the back, lightly and perhaps a little passive-aggressively, and we went our separate ways.

Hans Moravec (the Carnegie Mellon robotics professor who outlined a speculative procedure for transferring the material of human brains to machines) projects a future in which, "by performing better and cheaper, the robots will displace humans from essential roles." Soon after that, he writes, "they could displace us from existence." But as a transhumanist, Moravec doesn't see this as something to be feared, or even necessarily avoided; because these robots will be our evolutionary heirs, our "mind children," as he puts it, "built in our image and our likeness, ourselves in more potent and efficient form. Like biological children of previous generations, they will embody humanity's best chance for a long-term future. It behooves us to give them every advantage and to bow out when we can no longer contribute."

There is, obviously, something about the idea of intelligent robots that frightens and titillates us, that fuels our feverish visions of omnipotence and obsolescence. The technological imagination projects a fantasy of godhood, with its attendant Promethean anxieties, onto the figure of the automaton. A few days after I returned from Pomona, I read that Steve Wozniak, the cofounder of Apple, had spoken at a conference about his conviction that humans were destined to become the pets of superintelligent robots. But this, he stressed, would not necessarily be an especially undesirable outcome. "It's actually going to turn out really good for humans," he said. Robots "will be so smart by then that they'll know they have to keep nature, and humans are part of nature." The robots, he believed, would treat us with respect and kindness, with a patrician generosity, because we humans were "the gods originally."

It would seem to be among the oldest collective fantasies of our species, this fantasy of creation. It would seem to be part of us, a thing we carry with us across cultures and centuries, a dream of burnished hardware that replicates our bodies and acts in accordance with our desires. Frustrated gods that we are, we have always dreamt of creating machines in our own image, and of re-creating ourselves in the image of these machines.

Hellenic mythology had its automata, its living statues. The artificer Daedalus, remembered mostly for his disastrous efforts at human enhancement (labyrinth, waxen wings, tragic but morally instructive drowning), was also a maker of mechanical men, animated effigies capable of walking, speaking, weeping. Hephaestus, the blacksmith god of fire and metal and technology, constructed a bronze giant named Talos, to protect Europa (whom his father, Zeus, had abducted) from any further abductions.

Medieval alchemists were obsessed with the idea of creating men from scratch, believing it was possible to bring into being tiny humanoid creatures called homunculi; this they insisted could be done through arcane practices involving such diverse substances as cow's wombs, sulfur, magnets, animal blood, and locally sourced semen (preferably the alchemist's own).*

Saint Albertus Magnus, a thirteenth-century Bavarian bishop, was said to have constructed a metal statue with the power of reason and speech. According to popular accounts from the time, this alchemical AI, which Albertus referred to as his "android," met a violent end at the hands of a young Saint Thomas Aquinas, who was then a student of Albertus, and had serious issues with the android's incessant chatter

* The Catholic Church, famously, was not a fan of the whole alchemy scene. This had largely to do with the perception that Satan's fingerprints were all over the practice, with its herbs and sulfur and general aura of magic. But the fact that the work itself must have involved a nontrivial amount of wanking can't have helped, either.

and, even more problematically, its obvious origins in some kind of diabolical covenant.

In Europe, with the increasing popularity of clockwork during the Renaissance, and as the Enlightenment project supposedly cleared the mists of occult superstition from the field of science, there was a surge of interest in automata. In the 1490s, in an expansion of his own anatomical studies likely inspired by reading of the ancient Greek automata, Leonardo da Vinci designed and built a robotic knight. This automaton, often considered the world's first humanoid robot, was a suit of armor animated by internal cables and pulleys and gears. The knight, built for display at the home of Ludovico Sforza, the Milanese duke who had commissioned *The Last Supper,* was capable of a range of movements, including sitting, standing, waving, and simulating speech by moving its armored jaw.

Descartes' *Treatise on Man*—which he never published in his lifetime for fear of the Church's reaction to its central thesis—is predicated on the idea that our bodies are essentially machines, moving statues of flesh and bone animated by a divine infusion of spirit or soul. Part one, entitled "On the Machine of the Body," draws an explicit analogy between the clockwork mechanisms so popular at the time and the inner operations of the human body. "We see clocks, artificial fountains, mills, and other similar machines which, even though they are only made by men, have the power to move of their own accord in various ways. And, as I am supposing that this machine is made by God, I think you will agree that it is capable of a greater variety of movements than I could possibly imagine in it, and that it exhibits a greater ingenuity than I could possibly ascribe to it." Descartes wanted us to consider that everything we are—all our "functions," including "passion, memory and imagination"—follow "from the mere arrangement of the machine's organs every bit as naturally as the movements of a clock or other automaton follow from the arrangement of its counter-weights and wheels."

The *Treatise* is a weird and vaguely disturbing text, more for how it is written than for its mechanistic message. It is a work less of philosophy than straightforward anatomy, which reads like a kind of technological primer. Descartes' insistence on repeatedly referring to the body and its constituent parts as "this machine" has a powerfully estranging effect; reading it, you begin to feel a growing distance from your own body, this complex edifice of interconnected and autonomous systems—this soft machine within which you yourself, the impalpable reader of the *Treatise,* reside and hold sway. That this idea seems both utterly absurd and utterly familiar is a testament to the extent to which Cartesian dualism has, over the centuries, become a rigid orthotic structure around our relationships with our bodies. (The fact that a distinction between "us" and "our bodies" is even intelligible seems itself largely a result of his philosophy's despotic influence over how we think about these machines of ours.)

Descartes was also subject to what you'd imagine to be a peculiarly modern, or postmodern, preoccupation: the anxious imagination of *actual* machines that might pass themselves off as human. In his *Discourse on Method,* the famously rigorous austerity of his doubt is brought to bear on the contemporary vogue for automata, and its epistemological implications. Gazing out his window, he draws our attention to the people passing below. "In this case, I do not fail to say that I see the men themselves," he writes, "and yet, what do I see from the window beyond hats and cloaks that might cover artificial machines, whose motions might be determined by springs." If you're going to take your doubts seriously, in other words, if you're going to have the courage of your solipsism, what grounds do you have for believing that the man on the street—or for that matter the other dude driving your Uber—is not literally a machine, a replicant passing itself off as human?

In 1747, about a century after Descartes' death, the French physician Julien Offray de La Mettrie wrote a highly controversial pamphlet called *L'Homme Machine,* which literally translates as "Man a Machine." In it, La Mettrie takes a radical step beyond Descartes by jettisoning entirely the notion of the soul, and portraying the human creature as no different in kind to the animals Descartes had presented to the world as mere machines. For him, the human body was "a self-winding machine, a living representation of perpetual motion."

La Mettrie had been influenced by seeing the exhibited automata of the inventor Jacques de Vaucanson, whose most famous work was a mechanical duck that, when fed grain, appeared to have the ability of metabolizing and then defecating it. ("Without the shitting duck of Vaucanson," Voltaire sharply observed, "we would have nothing to remind us of the glory of France.") Vaucanson also made human automata, although these were charged not with the production of feces, but with more genteel labors like the playing of flutes and the banging of tambourines.

It was through the popularity of Vaucanson's contrivances that the term "android" became established. The first volume of Diderot and d'Alembert's *Encyclopédie* contained a lengthy and detailed description of Vaucanson's automated flautist, in an entry entitled "Androïde," which referred to "an automaton in human form, which, by means of certain well-positioned strings, etc. performs certain functions which externally resemble those of man."

In *L'Homme Machine,* La Mettrie raises the specter of an automaton capable of more than mere parlor tricks. "If it took more instruments," he writes, "more cogs, more springs, to show the movement of the planets than to show or tell the time, if it took Vaucanson more artistry to make his flautist than his duck, he would have needed even more to make a speaking machine, which can no longer be considered impossible, particularly at the hands of a new Prometheus."

In 1898, when the power of the U.S. Navy was being tested in the Caribbean and Pacific during the Spanish-American War, the inventor Nikola Tesla displayed a new device at an electrical exhibition at New York's Madison Square Garden. This was a miniature iron boat, which Tesla had placed in a large vat of water, and equipped with a mast for the reception of radio waves, allowing him to direct its movements from the opposite end of the arena with a wireless controller. The demonstration stirred considerable public excitement, and Tesla and his autonomous boat made the front pages of national newspapers. Given the events of the time, the device was inevitably interpreted as a great leap forward in the technology of naval warfare. But like so many scientists whose innovations have refined the machineries of slaughter, Tesla was personally opposed to the forces of nationalism and militarism (if only passively so). According to *Prodigal Genius*, a 1944 biography by John O'Neill, when a student suggested that the boat might prove extremely useful if its hull was packed with dynamite and torpedoes to be remotely detonated, Tesla snapped: "You do not see there a wireless torpedo; you see the first of a race of robots, mechanical men which will do the laborious work of the human race."

Tesla was convinced that the development of this "race of robots" would have a transformative influence on how humans lived and worked, on how they waged war. "This evolution," he wrote in 1900, "will bring more and more into prominence a machine or mechanism with the fewest individuals as an element of warfare.... Greatest possible speed and maximum rate of energy delivery by the war apparatus will be the main object. The loss of life will become smaller."

Writing in June 1900 about his ambition to create functioning humanoid robots, Tesla echoes Descartes and La Mettrie in invoking his own sense of himself as a mechanical instrument:

> I have by every thought and act of mine, demonstrated, and do so daily, to my absolute satisfaction that I am an automaton endowed with power of movement, which merely responds to external stimuli beating upon my sense organs, and thinks and moves accordingly.
>
> With these experiences it was only natural that, long ago, I conceived the idea of constructing an automaton which would mechanically represent me, and which would respond, as I do

> myself, but, of course, in a much more primitive manner, to external influences.

This machine, he reasoned, "would perform its movements in the manner of a living being, for it would have all the chief elements of the same." To the problem of this machine lacking an "element" of mind, Tesla proposed the solution of letting it borrow his own. "This element," he writes, "I could easily embody in it by conveying to it my own intelligence, my own understanding." The idea here was that he would control the machine using the precise method he had used with the boat. This he gave the inelegant name "telautomatics," by which he meant "the art of controlling the movements and operations of distant automatons."

But he was convinced that it would be possible to create automata not merely with borrowed minds, but with the ability to think for themselves. As he put it in an unpublished statement fifteen years later, "Teleautomata will be ultimately produced, capable of acting as if possessed of their own intelligence and their advent will create a revolution."

Over the two days I spent at the Fairplex, I saw things that caused me to reflect upon whether such a revolution might be close at hand. In a more or less explicit sense, the whole premise of the event was that these automata would, sooner or later, stand in place of our bodies, these machines of bone and sinew and flesh. I saw a bomb-disposal robot, its pincered arms moving in perfect synchrony with those of a man standing behind it, opening a zippered canvas bag and plucking from within it plastic-wrapped candy, which it handed to passersby—as powerful an example of Tesla's teleautomatics, in its way, as the more complex humanoid marionettes competing in the arena. Tesla's idea of a race of robots "doing the laborious work of the human race" was clearly some distance from realization, but there didn't seem to be any doubt that this was what capitalism's most advanced engines were driving toward. A solid indicator of this trend was offered, as it happened, by a business named after Tesla himself: the Silicon Valley electric car company Tesla Motors, whose production line was almost entirely roboticized, and whose CEO, Elon Musk—the same Elon Musk who was so publicly terrified by the prospect of artificial superintelligence—had recently announced the company's plans to develop its own self-driving system within three to five years.

Although I had not beheld him with my own human eyes, I understood that Musk had come to the Fairplex that weekend to observe the robots and meet with their engineers. And I understood, too, that Google's cofounder Larry Page, a Singulatarian of note, had descended from the summit of Mountain View to be among these machines, in whose future his own company had invested considerable money. In 2013, Google had paid half a billion dollars for Boston Dynamics, whose menagerie of uncanny creatures—BigDog, Cheetah, Sand Flea, LittleDog—had been created largely with DARPA funding, and whose Atlas robot was being used as hardware by several of the teams here in Pomona.

A few hundred yards from the racetrack, in the massive hangarlike building from which the robots were directed by their engineers, a squad of Boston Dynamics technicians was also on hand to tend to the contusions and malfunctions of the Atlas humanoids.

Boston Dynamics, with its weird techno-fauna, was itself a hybrid specimen of the relationship between the Pentagon and Silicon Valley; its machines were the unnatural creatures of a new military-industrial complex. Google's links with DARPA were numerous and far-reaching. DARPA's previous director, Regina Dugan, for instance, had left her government job to work at Google's headquarters in Mountain View, where she now leads something called the Advanced Technology and Projects Team.

For a while now, I had been fascinated by the creatures produced by this robotics firm, founded in the early 1990s by Marc Raibert (a former colleague of Hans Moravec at Carnegie Mellon's Robotics Institute). Over the last couple of years, I had repeatedly and compulsively watched the succession of YouTube videos released by the company, containing footage of their latest ingenious automata. And I found something subtly unsettling in these robots, in their simultaneous remoteness from and proximity to recognizable forms of biological life. Looking at BigDog, for instance, skittering with blind insectile relentlessness over a patch of ice, or WildCat, with its uncanny hydraulic dressage, I would feel a pleasurable thrill of dread—an instinctive terror of predation, perhaps—compounded with the knowledge that these robots had been created with Pentagon funding, and had become by acquisition the creatures of the world's most powerful technology corporation.

The rhetoric of Silicon Valley's geek establishment is steeped in a diluted solution of countercultural idealism—changing the world, making things better, disrupting old orders, and so forth—but its roots are deep in the blood-rich soil of war. As the writer Rebecca Solnit puts it, "the story Silicon Valley rarely tells about itself has to do with dollar signs and weapons systems."

Hewlett-Packard, the valley's first major success, was a military contractor whose cofounder David Packard served as deputy secretary of defense during the Nixon administration. His most significant contribution during his term of office, Solnit points out, "was a paper about overriding the laws preventing the imposition of martial law."

I was aware that there was something unreasonable, even slightly hysterical, in my reaction to Boston Dynamics' menagerie of humanoids and techno-animals, some half-gleeful indulgence of a paranoid tendency, but I could not on that account disregard that reaction. At a subcortical level, I rejected these creatures and what they represented; some primitive, human part of me wanted to smash them with a hammer just as the young Thomas Aquinas destroyed the automaton of Albertus Magnus. I was subject, in other words, to an obscure but insistent sense of their diabolical provenance and intent.

And yet I felt the whole idea of political paranoia to be increasingly an anachronism, a nostalgic and basically meaningless gesture toward a twentieth-century dispensation in which it was possible merely to *suspect* the iniquitous designs and collusions of government and capital. To be paranoid now—paranoid, that is, as opposed to minimally aware of what was going on—was to be deluded in a manner that had become almost whimsical, like those fond sentimentalists who comfort themselves with folktales about secret world governments and shapeshifting lizards and Illuminati bloodlines, and to whom the only reasonable response was to say, "Look, pal, you're overthinking it, have you looked into this whole *capitalism* deal?" It was out in the open now, the truth about these things—or enough of it to be getting along with.

The May 1924 issue of the American popular technology magazine *Science and Invention* featured a cover illustration of a colossal red robot—a thing like an outsized hot water cylinder with articulated legs and caterpillar treads for feet and, instead of hands, twin whirring circular blurs of truncheons. The shining yellow lamps of its eyes stare down at a scattering crowd of men, their peaked caps flying from their heads, eyes abulge with terror as they flee the robotic assault. The article inside, entitled "Distant Control by Radio Makes Mechanical Cop Possible," describes this imagined law enforcement device in almost obscene detail: the stabilizing gyroscopes of its thighs, the radio control cabinet and gasoline tank in its thorax, the modest phallic presence of its tear gas duct, the anal conduit of its rear engine exhaust. A further illustration shows a towering phalanx of these robot cops driving back a crowd of protesting workers, against a desolate background of chimney stacks, louring smoke, dark satanic mills. "Such a machine," we are assured, "would seem to be exceedingly valuable to disperse mobs, or for war purposes or even for industrial purposes. For fighting mobs use is made of tear gas which is stored in a tank under pressure and which alone will quickly displace a mob if necessary. The arms are provided with rotating discs which carry lead balls on flexible leads. These act as police clubs in action."*

This bare fascist fantasy, revealing for all its absurdity, depicts a violent machinery of the state protecting the interests of capital against the assembled bodies of the laboring class, with their regulable human wills, their frangible skulls beneath undoffed caps. It's as blunt an illustration as I've seen of the prewar terror of organized labor: an inverted Frankenstein scenario, whereby the monstrous body of the automaton, a looming literalization of Hobbes's "Artificiall Man," is enlisted into the stern husbandry of ideological order. As the French philosopher Grégoire Chamayou puts it in his book *Drone Theory*, the dream represented by the "mechanical cop" is "to construct a bodiless force, a political body without human organs, replacing the old regimented bodies of subjects by mechanical instruments that would, if possible, become its sole agents."

Even as I cheered on the robots competing at the DARPA Robotics Challenge, and even as I laughed indulgently at their crude pratfalls, something of this technological unease clung to me through my time in Pomona, some sense that I was watching the first staggering movements toward an unmanned future.

As I took my leave of these machines, the mind-children of my fellow humans, and as I made my way out of the grandstand toward the Uber driver whose approach I was tracking on my iPhone screen, I found myself suddenly, intimately conscious of the mechanical

* The author of this article—and the publisher of the magazine in which it appeared—was Hugo Gernsback, the Luxembourgian-American inventor and entrepreneur who is often credited as a founding figure of modern sci-fi, mostly for having gone on to publish the first-ever dedicated sci-fi journal, *Amazing Stories*. The Hugos, the annual awards for outstanding achievement presented at the World Science Fiction Convention, are named after him. Like many a successful businessman then and now, Gernsback clearly did not have a lot of time for unions, whose members he seemed to want to be fucked in the eye by tear-gas-ejaculating police penises.

nature of my movements, of the articulated pendula of my legs, with their ball joints and adductors and extensors, and I felt for a moment as though no interior volition was at work here, as though this object in motion, feeling these things, was merely a component in some vast and unrevealed pattern, some controlled system that included the Uber driver, the advancing car, the highway network of Greater Los Angeles, the images representing these phenomena on the smartphone screen, the eyes watching that screen, the information, the code, and the world itself, among other things.

It occurred to me, not for the first time, that I might be losing my mind, that I might be succumbing to some bizarre delusion, brought on by excessive exposure to humanoid machines and to mechanistic ideas of the human being, that I myself was a machine, or a subordinate mechanism in a vast universal contraption of every existing thing. It was a delirium, or a truth, that would gather momentum, and find a sort of uncanny external reflection in the machinic humans, the self-proclaimed cyborgs, I was about to encounter.

Biology and Its Discontents

THE OLD STEUBENVILLE Pike is a narrow stretch of country road just off the freeway between downtown Pittsburgh and the airport. A short distance along this route, there's a small motel that's been abandoned since the 1950s, its cracked windows and wooden doors half concealed by a profusion of vegetation: a rotting avatar of definitive Americana in the midst of nature's slow and implacable reclamation. Right next door is a small wooden house with a couple of canvas hammocks hanging from its front porch.

If you were passing by here, maybe on the way to pick up a crate of beer at the drive-thru liquor wholesaler just down the road, you might notice some people lounging in those hammocks, leaning against the screen door. And if you did notice them, you probably wouldn't think there was anything much going on with them; you'd probably assume they were just a bunch of young wasters sitting around on a porch, smoking, shooting the Western Pennsylvania breeze. You'd have no reason, certainly, to believe that they were cyborgs, or that they thought of themselves as such. You'd have no reason to believe that they'd just come up for a breather from the basement of that house, where they'd been tinkering with homemade technologies for transcending the limitations of the human animal.

Let me fill you in, briefly, about that basement, where I spent some disorienting afternoons and evenings with these people, these cyborgs, in the dying days of the summer of 2015. It didn't look like the kind of place where the future, or *a* future, was being created. It could definitely have done with a good cleaning, for one thing. There was stuff everywhere, a grimy miscellany of disjecta: disemboweled hard drives, decommissioned monitors, empty beer bottles, cardboard boxes, forsaken workout equipment coated with a velvet layer of dust. When I arrived on my first evening there, the occupants of this basement were unfurling a new plastic banner they'd just taken delivery of; in a spirit of corporate pride, they were tacking it to a wall over a long desk cluttered with an array of devices—laptops, semiconductors, batteries, wires, oscilloscopes. The banner bore the words "GRINDHOUSE WETWARE" in a stockily futuristic font, and a red and white stylized image of silicon-chip circuitry in the shape of a human brain.

Grindhouse Wetware is, as the company's website explains, a team of people working toward the goal of "augmenting humanity using safe, affordable, open source technology." Their devices are designed for subdermal implantation, and intended to enhance the sensory and informational capacities of the human body. Grindhouse is the most prominent group within what is known as the grinder scene, a mostly online community of biohackers, or "practical transhumanists." These are people who don't want to wait around for the Singularity to happen, or for artificial superintelligence to finally materialize and subsume the informational content of their human minds, their wetware. With the means currently at hand, they are doing what they can to merge with technology right now.

The company, as it happened, had just had a nontrivial infusion of cash, and there was a palpable sense of relief, of accomplishment, in the air. As of this evening, the cyborg future was ten grand closer. A check had just hit the company's bank account, payment for a recent speaking engagement in Berlin by one Tim Cannon, Grindhouse's chief information officer and de facto leader, out of whose basement the whole operation was run.

Earlier that evening, I had met up with Tim and a couple of his

Grindhouse colleagues at a place called TechShop, a maker space in Pittsburgh's Oakland neighborhood, where he was taking part in a panel discussion that was being recorded for an NPR show. It was our first encounter in the flesh, after almost a year of emails and Skype chats, most of which had been mediated by Grindhouse's publicity director, Ryan O'Shea. Ryan had come along for the talk, as had their young colleague Marlo Webber, a gifted self-taught electrical engineer who had recently moved from northeastern Australia in order to work with him. He'd been crashing at Tim's place since he got to Pittsburgh; the plan was that the company would eventually be in a position to pay him a wage sufficient to make him eligible for a work visa.

These gentlemen didn't look like cyborgs, although I suppose that raises the question of what you'd expect a cyborg to look like. They didn't especially look like geeks, is maybe what I'm saying. Ryan looked more or less how you'd expect a guy to look who had a day job in independent film production, and who'd previously worked as a congressman's aide on Capitol Hill: neat blond hair, black-framed Ray-Ban prescription glasses, beige slacks, checked shirt—his style was situated somewhere in the disputed middle ground between hipster and preppy. Marlo was small, slight, in jeans and a black denim shirt; he had the face of a wayward teen, a permanent half smile, like he was relishing some private absurdity, weighing up whether to disburden himself of a smart-ass remark that had just occurred to him.

And Tim: given that his whole deal was radical self-transformation, Tim looked very much like a guy who'd worked out an aesthetic for himself at sixteen and basically butched it out since the late 1990s. He wore a black flat cap, a Grindhouse T-shirt, chunky skater shoes, and green cargo shorts that exposed a tattoo on his right calf depicting a cartoon punk (Mohawk, Dead Kennedys shirt) holding a gun to his own head. Another large tattoo, on the pale underside of his left arm, depicted a DNA double helix surrounded by a circular cog. This pictorial representation of Tim's mechanistic view of *Homo sapiens*—a grinding of the human code—was, literally, underscored by an impressively lurid scar, gnarled and thick-grained as bark. This was the result of a device called Circadia he'd had implanted for three months of last year; it took various biometric measurements from his body and uploaded them via Bluetooth to his phone, and thereby to the Internet, at five-second intervals—adjusting for good measure the thermostat of his house's central heating in accordance with his body temperature.

If you'd met Tim back then, you'd have found it difficult to ignore the protrusion about the size of a pack of playing cards bulging beneath the ventral surface of his forearm. You might have felt faint or nauseated even looking at it, this spectacle of techno-penetration, this violent machining of the flesh. The device's insertion required a long incision, then a lifting away of the upper layer of skin from the fatty tissues beneath to create a yawning orifice, then a penetration of the body with the device, before a final suturing of the wound, a stretching and closure of the meat over the machine. Because no medical professional could perform the procedure without losing their license, the whole thing was done in Berlin by a body-modification "flesh engineer," and it was performed in a manner Tim referred to as "raw dog," meaning without the benefit of anesthesia.

"That was a rough ninety days," he said, "I'll tell you that much."

We were lounging around in armchairs outside the room where the NPR people were setting up for the panel discussion, putting out popcorn and spring water and craft beer.

"In the first couple of weeks, there was a lot of fluid buildup, and that had to be drained regularly. I also had to take medication to stop my body from rejecting the implant. I was in a state of constant paranoia. I'd feel a tingling in my head, and I'd think it was my brain being poisoned by the battery leaking into my bloodstream or some shit. Then I'd sneeze, and I'd be like, oh, okay, I just needed to sneeze."

People would ask him why Circadia was so large, and he would tell them this was because they weren't trying to make it small. It was a proof of concept, an experiment to ascertain whether the technology functioned as required inside the body. And it did: Tim's mortal terror aside, it functioned just fine. Now they were working on a newer and more compact version, requiring a less grotesque and gratuitous transgression of the membrane between human and machine.

Tim told me about his grueling-sounding daily routine, which involved a day job as a programmer with a software contracting agency, and nights working on Grindhouse stuff down in the basement. He also had a son and daughter, aged nine and eleven respectively, over whose custody he was involved in an intense and protracted legal struggle with his estranged ex. None of this left a lot of time for sleep, and so he'd implemented a system of polyphasic sleeping, whereby he took two twenty-minute naps during the day, and then went into shutdown mode for a total of three hours every night, generally between 1 and 4 a.m.

It was all about systems, he said, about understanding and manipulating them: the system of the day, the system of the body, the system of a life.

A middle-aged woman in high-waisted trousers and sandals wandered over to where we were sitting. She had a scrubbed and oddly blank appearance, and her hair was pulled back austerely from her face. She introduced herself as one of Tim's two fellow interlocutors on the panel. Her name was Anne Wright; she was a professor at Carnegie Mellon, and she was heavily involved in the Quantified Self movement, whose adherents used technology to track and analyze data about their everyday lives. Tim told her he'd dabbled in QS himself, that he'd recently bought a wearable gadget that tracked his every movement and uploaded the data to the cloud for later analysis. He was interested in the whole Quantified Self thing, he said, but had his reservations.

"It's really a matter of gathering as much data as possible about your life," she told him, "and figuring out how you can use that data to optimize yourself as a person."

"Right," said Tim. "Although I'd like to take the 'person' term completely out of the equation. People really suck at decisions. It's like the whole self-driving car thing. People are like, 'Oh, you can't take humans out of the loop, I'm a human and I'm an *awesome* driver.' And I'm like, no, man, you're not an awesome driver. You're a monkey, and monkeys suck at making decisions."

Anne emitted a perfunctory laugh. She seemed uncomfortable, and I wondered whether this discomfort was a reaction to the way in which Tim's language seemed to lay bare the mechanistic principles of the QS movement, its view of the self as reducible to a set of facts and statistics that could be interpreted, and whose interpretation thereby informed the activity of the self, and thereby the generation of further data—the human being as a feedback loop of input and output.

"Far as I'm concerned," Tim went on, "there is no amount of optimization of this barely evolved chimp that is worthwhile. We just don't have the hardware to be ethical, to be the things we say we want to be. The hardware we do have is really great for, you know, cracking open skulls on the African savanna, but not much use for the world we live in now. We need to change the hardware."

It was Tim's rhetorical custom, in common with many transhumanists, to make frequent allusions to the African savanna. We were a long way from the world for which we evolved, was the general point.

"This guy's like some kind of quote-generating machine," I said to Marlo, as we waited for the panel to start. I was kneading my wrist, massaging its crude technology of ligament and cartilage, the seized carpal machinery beneath its casing of skin.

"My writing hand's fucked already," I said. "Maybe you guys could fix me up with some kind of transcription upgrade."

Marlo chuckled, and showed me an RFID chip he'd implanted in the back of his own hand, probing it back and forth through the thin layer of flesh with an index finger. It was roughly the size and shape of a paracetamol capsule. In theory, it enabled him to wave his hand and unlock the front door of HackPittsburgh, the laboratory space downtown where they sometimes worked when they needed more high-grade equipment, but as a new employee he didn't have clearance, so it basically just sat there, a dormant cell of technology awaiting its commands.

The panel discussion was entitled "Borg in the USA: Cyborgs and Public Policy in the Digital Age," and, with Anne Wright, the other speaker was an elegantly dressed man named Witold "Vic" Walczak, legal director of the ACLU for the state of Pennsylvania. The moderator, an NPR presenter called Josh Raulerson, introduced Tim by announcing to the room that "there are, it's safe to say, literal cyborgs among us," and then looked at Tim and asked whether this was in fact safe to say, or whether he'd perhaps used an inappropriate term.

"It's as good a label as anything else," said Tim, and shrugged.

There was some back-and-forth on the topic of Big Data, and of contemporary humanity as a collection of nodes through which information is channeled. Anne talked at length about her discomfort with corporations using this collected information about her to predict what she might want to purchase, or where she might want to travel to. Tim, for his part, said that there was a difference between using people and utilizing them. He didn't understand why everyone was so precious about being predicted.

"I think it offends people's belief that they're these unique little flowers. But we're animals, and animals have patterns of behavior. We're offended by any suggestion that we're predictable."

"I'm not predictable," said Anne, who had, predictably, taken offense.

"Everyone is predictable," said Tim, "given a high-enough degree of information, and powerful-enough processing."

Here, Anne invoked the academic-sounding concept of "emplotment," whereby a person became "emplotted" in some external design, used in a narrative other than their own. "There's something wrong with this putting together of a person's pattern," she said. "It makes us into a character in someone else's plot."

I noted here that Marlo, who had been availing himself remorselessly of the complimentary pale ales, was shaking his head in weary disputation.

"If a computer's able to predict to ninety-nine-point-nine-nine-nine percent accuracy from your purchases or your search engine activity that you're pregnant," said Tim, "that's not 'emplotment,' that's just a fact. We are deterministic mechanisms. The problem is, most people make the mistake of anthropomorphizing themselves."

This latter aphorism landed slow and unsteady, drawing laughter from perhaps only half of the fifty or so people in the room—a strangled laughter, an uneasy laughter, a laughter of uncertain subjects.

Our need for privacy, said Tim, arises out of our primitive animal nature. If we had more advanced brains, we wouldn't do things that required the screen of privacy. The solution, he said, was to go inside the brain and destroy vestigial behaviors that are no longer useful, because evolution was just not happening fast enough.

"I mean, we are breeding at an unsustainable rate here, devouring all our resources. Our libido is calibrated for the Ice Age, where one in four of us died in birth and took the mother with us. That's no longer the case. And yet every one of us in this room is *very* interested in fucking. Right?"

There was another ripple of nervous laughter. Josh Raulerson smiled queasily at the audience, and Ryan shifted in his seat.

"Man, I hope this isn't live," said Tim. "*Is* it live? Because I've just been saying whatever I feel like saying."

The thing about listening to Tim and his fellow grinders, the activity in which I was largely occupied while I was in Pittsburgh, is that their rhetoric is so forceful that it pushes you to adopt defenses of positions you're not sure you hold. Their whole ethos, in one sense, is such a radical extrapolation of the classically American belief in self-betterment that it obliterates the idea of the self entirely. It's liberal humanism forced to the coldest outer limits of its own paradoxical implications: If we truly want to be better than we are—more moral, more in control of ourselves and our destinies—we need to drop the pretense that we are anything more than biological machines, driven by evolutionary imperatives that have no place in the overall picture of the kind of world we say we want to create. If we want to be more than mere animals, we need to embrace technology's potential to make us machines.

The idea of the cyborg is mostly associated with science fiction—with Philip K. Dick and William Gibson, with *RoboCop* and *The Six Million Dollar Man*—but its origins are in the postwar field of cybernetics, which its founder, Norbert Wiener, defined as "the entire field of control and communication theory, whether in the machine or in the animal." In the posthumanist vision of cybernetics, human beings were not individuals acting autonomously toward their own ends, free agents in pursuit of their destinies, but machines acting within the deterministic logic of larger machines, biological components of vast and complex systems. And what linked the elements in these systems was information. The key idea of cybernetics was the notion of the "feedback loop," whereby a component in a system—for instance, a human—receives information about its environment, and reacts to that information, changing the environment and thereby the subsequent information it receives. (The Quantified Self movement is, in this sense, deeply embedded in the cybernetic worldview.) Where energy, its transformations and transfers, had previously been seen as the fundamental building block of the universe, information was now the unit of universal exchange. In cybernetics, everything is a technology: animals and plants and computers were all essentially the same type of thing, carrying out the same type of process.

The term cyborg, which means "cybernetic organism," was first used in a 1960 scientific paper called "Cyborgs in Space," published in the journal *Astronautics* by the neurophysiologist Manfred Clynes and the physician Nathan Kline. The article begins with the fairly uncontroversial assertion that the human body is constitutionally unsuited to space exploration, and goes on to suggest that it would therefore be beneficial to integrate into the bodies of astronauts such technologies as would allow them to function as self-sustaining systems in hostile extraterrestrial environments. "For the exogenously extended organization complex functioning as an integrated homeostatic system unconsciously," they write, "we propose the term 'Cyborg.' The Cyborg deliberately incorporates exogenous components extending the self-regulatory control function of the organism in order to adapt it to new environments."

And so the cyborg arose as a Cold War phantasm, a dreamlike intensification of American capitalism's ideals of efficiency and self-reliance and technological mastery. One of the variously conflicting and interlocking definitions of the cyborg Donna Haraway offers in her essay "The Cyborg Manifesto" was as "the awful apocalyptic telos of the 'West's' escalating dominations of abstract individuation, an ultimate self untied at last from all dependency, a man in space." It was also a kind of reductio ad absurdum of the mechanistic and militaristic vision of the human body and brain: the cyborg was the human not simply as machine, but specifically as war machine—a human body and mind in a symbiotic feedback loop with the information systems of modern warfare.

The American government has, unsurprisingly, shown a longstanding interest in the idea of merging humans with machinery for military purposes. In 1999, DARPA began awarding grants for "biohybrid" research programs, the objective of which was to create creature-machine crossbreeds. That was the year the agency created its Defense Sciences Office, and hired a former McDonald's executive and venture capitalist named Michael Goldblatt as director. Goldblatt was convinced, as he put it in an interview, that the "next frontier was inside of our own selves," and that human beings could be "the first species to control evolution." As Annie Jacobsen puts it in *The Pentagon's Brain,* her relentlessly approving book about DARPA, Goldblatt was "a pioneer in military-based transhumanism—the notion that man can and will alter the human condition fundamentally by augmenting humans with machines and other means."

The agency's funded programs began producing various nightmarish chimeras: rats whose movements could be controlled from laptops via electrodes implanted in their medial forebrains, hawk moths with semiconductors implanted at pupal stage so that the technology became part of their adult development. By getting in on the ground floor of the insect's metamorphic tissue development, Jacobsen writes, the scientists "were able to create a steerable cyborg, part insect, part machine." (Wiener's coinage "cybernetics" arose initially from the Greek word *kubernan,* meaning "to steer.")

As DSO director, Goldblatt was more or less frank about the aspiration to create human-machine hybrids, super-soldiers who would be built to withstand and thrive in the extreme conditions of combat. In a statement to his program managers not long after DARPA hired him, Goldblatt insisted that "soldiers having no physical, physiological, or cognitive limitation will be key to survival and operational dominance in the future." Areas of experimentation included pain vaccines, chemical compounds through which injured soldiers could go into a kind of "suspended animation" until medical aid arrived, and a "Continually Assisted Performance" program, which sought to create a "24/7 soldier" who would gain the edge on enemy combatants through never needing to sleep.

Brain-machine interfaces became a major area of interest for DARPA around the turn of the century, and remain a significant target of funding. The idea is to allow soldiers to communicate and control purely by thought. "Imagine a time," as the DSO's Eric Eisenstadt put it, "when the human brain has its own wireless modem so that instead of acting on thoughts, warfighters have thoughts that act."

All of this seemed to further pierce the veneer of cheerful humanitarianism that had surrounded the Robotics Challenge in Pomona. DARPA's interest in technology was, it seemed clear, always an interest in the methodology of efficient violence.

—

The grinder movement is characterized by both an internalization and a subversion of this cybernetic ideal. Grinders want the same things as DARPA, but for individualistic reasons; the aim, in this sense, is a kind of personalized military-industrial complex. "The main trouble with cyborgs," as Haraway puts it, "is that they are the illegitimate offspring of militarism and patriarchal capitalism, not to mention state socialism. But illegitimate offspring are often extremely unfaithful to their origins. Their fathers, after all, are inessential."

There is a strong element of performance to the grinder movement; Tim's implanting of a gigantic biometric measuring device in his arm, to take an unignorably bulging example, is as much a provocative gesture as anything else. In this sense, one obvious forebear of the movement is the Australian performance artist Stelarc, whose work since the 1970s has been increasingly extreme in its obliteration of the boundaries between technology and flesh. For his piece *Ping Body*, he attached electrodes to his muscles, which allowed his body movements to be controlled by remote users over the Internet. *Ear on Arm*, a project he began in 2006, involved the use of cell cultivation and surgical construction to create a prosthetic ear on his left forearm, with the intention of connecting it to the Internet so that it could be employed as a "remote listening device" for people in distant locations. Stelarc's whole artistic project is an explicitly transhumanist one: a series of provocative gestures intended to represent the body as a technology in need of updating for the information age. As he put it in a statement that directly echoed the original invocation of the cyborg by Clynes and Kline: "It is time to question whether a bipedal, breathing body with binocular vision and a 1400cc brain is an adequate biological form. It cannot cope with the quantity, complexity and quality of information it has accumulated; it is intimidated by the precision, speed and power of technology and it is biologically ill-equipped to cope with its new extraterrestrial environment." The body, for Stelarc—the poor, bare, forked animal that we are—is an obsolete technology. Flesh is a dead format.

What would it mean to think of yourself—you, personally—as a cyborg? In a certain sense, the idea of the cyborg is no more or less than a particular way of thinking about the human, a peculiarly modern picture of a person as a mechanism for the processing of information. Do you wear glasses? Do you wear orthotics in your shoes? Do you have a pacemaker fitted in your heart? Do you get a strange phantom limb sensation when you are for some reason denied access to your smartphone, when your battery's dead or your screen is smashed or you left the thing in your other jacket, and so you can't access some or other piece of crucial information, or you can't navigate via GPS, can't triangulate your location using a satellite orbiting the Earth? And are you therefore lost? Does that lostness, that loss, suggest a breakdown in the exogenously extended organization complex of your body and its supplementary technologies, a rupture in the integrated homeostatic system of yourself? If a cyborg is a human body augmented and extended by technology, is this not what we basically are anyway? Are we not, as they say in the philosophy racket, *always already* cyborgs? These aren't rhetorical questions. I'm genuinely asking here.

On my second day in Pittsburgh, I found myself with an afternoon to kill before taking a cab out to Tim's house to meet the grinders again. So I left my hotel downtown and walked toward the river, following on the screen of my phone the pulsing blue circle representing the location of my body in space as it inched downward through the empty grid of streets. In the basement of a museum dedicated to the work of one of the city's most famous citizens, Andy Warhol, I saw a poster bearing a monochrome image of the artist crouched over a silkscreen mesh, beneath which was printed a quotation: "The reason I'm painting this way is that I want to be a machine."

Later, as I lingered in the gift shop—which, more than ordinarily, in this particular museum seemed like the point of the place—I plucked from a shelf a paperback edition of the screenplay for the film *I Shot Andy Warhol*, in which Lili Taylor plays Valerie Solanas, the writer and former sex worker who attempted to murder Warhol in 1968. In the back of the book, the full text of Solanas's compellingly insane and disturbingly insightful *SCUM Manifesto* was reproduced. Flicking through it, I came across the following lines: "To call a man an animal is to flatter him; he's a machine, a walking dildo."

I replaced the book on the shelf and, neither flattered nor wounded, walked out of the museum and back across the river.

"People just routinely give themselves too much credit," Tim said.

The blades of a ceiling fan turned slowly above our heads, and through the screen door in Tim's kitchen the evening trill of cicadas could be heard, the clicking and whirring of countless machines connecting to the system of the night.

He said: "If you look at the evolution of the brain, the logic centers, they were growing at the same time as the creative centers were expanding. And that creates this really potent illusion that you're not just a bag of chemicals reacting to shit. Which is what you are."

Tim was leaning with his back to the sink over by the kitchen window, and on the wall behind his head were stenciled, in ornate script, the words "Live well, Love much, Laugh often." This sentiment sat oddly with its surroundings, and with what was under discussion among the bags of chemicals in the room. This flourish of interior decor, I surmised, was the work of Danielle, Tim's cheerfully reserved girlfriend, a Web developer who worked for the Pittsburgh Cultural Trust. Danielle did not herself have any special transhuman aspirations, but was open to the idea of maybe getting an implant at some point.

"Maybe down around here," she said, indicating her hip, "where it wouldn't be so visible."

Having lived with Tim for eight years now, she'd accommodated herself to the strangeness of his work and his lifestyle, to the extremity of his futurist vision. She'd been with him when he first decided he wanted to be a cyborg, when he announced to her that, as soon as such a procedure became available, he was planning to have his arm amputated and replaced by a technologically superior prosthesis.

They were in the car when he came out with it: once they started making prostheses that were superior to his natural limbs, he said, he would have no sentimental aversion to having those limbs removed and replaced with the more advanced technology. She was taken aback, even horrified at first, but she got used to the idea.

"If it makes him happy, I'm happy," she said. "It's whatever."

"People have this magic-in-the-meat mentality," said Tim. "People have this idea that because something is natural to our bodies, it's therefore somehow more real, more authentic."

This attitude, of which Danielle had cured herself over time, was, he insisted, irrational and sentimental. Every seven years, he said, our bodies were entirely replaced; and so he was, at the cellular level, not the person who had met Danielle eight years ago, and in another eight years he would be an entirely different person again: a different body, a different *thing*. Whether they were replaced by "natural" means, by the death and regeneration of cells, or by bionic prostheses, the arms with which Tim now embraced Danielle would no longer exist eight years from now.

I thought of Randal Koene, and of the idea that the physical forms we inhabited, the substrates of our existence, were purely contingent. I thought of Nate Soares saying there was "nothing special about carbon." I had no idea whether it was true or not, this idea that every cell in our bodies was replaced on a seven-year cycle, but if it was true, it seemed like it would be a potential propaganda victory for transhumanism, for substrate independence, for the Ship of Theseus view of whole brain emulation. It was a vertiginous thought: that the person who had first read about transhumanism in Dublin ten years ago had no material connection to the person now sitting in a living room in Pittsburgh talking to a transhumanist about how all the cells of the body were replaced on a seven-year cycle—because if there was no such material connection, how could either of those people be me, my "self"? And what was a "self" anyway? What was a person? Wasn't a person just a bunch of atoms, and wasn't an atom mostly emptiness—just a shell, containing a single nucleus floating in nothing? Wasn't a person therefore more or less a vacancy? I had begun questioning whether I could even be meaningfully said to exist when one of Tim's dogs wandered in from the back porch and took a frank and almost officious interest in my crotch; this I received as a sign that I did, in fact, exist—or a sign, at least, that I should move on to some other topic.*

I wanted to see some implants, and so Tim and Marlo took me down to the basement to look at what they were working on. The major project right now was a technology called Northstar, which Marlo described as his "baby." The current iteration was capable of detecting magnetic north, and lit up in response with red LEDs under the skin. A newer version, which Marlo was in the process of building, would incorporate gesture recognition, so that the implanted customer would be able to open their car door with, say, a circular motion of the palm, or start the engine by making a cruciform blessing in the air.

Neither Tim nor Marlo disagreed with my suggestion that, as impressively provocative as this kind of thing might be, it didn't exactly amount to a revolutionary intervention into the human condition. The idea of being able to unlock a car door with a gesture was, in itself, not much more than a gesture—a signaling toward some larger and more profound transformation. (It was also, if anything, less convenient than just using your car keys, given that using your car keys didn't involve unlicensed surgery.) But this, they insisted, was just the beginning. There was almost no limit to what you could achieve if you approached humanity as an engineering problem. And biology was the fundamental difficulty; the nature of the problem was nature itself.

* It turned out to be technically untrue. Although many of our organs regenerate their cells at various rates, there are cells in the body, such as those in the cerebral cortex, that are never replaced. This fact I was both relieved and obscurely disappointed to learn.

"We just shouldn't be in the biology game anymore," said Tim. "It's just not the right game for us, as a species. It requires too much wanton cruelty."

He was sitting cross-legged and barefoot on an office chair, taking hits from a modified vapebox, his face now and then obscured by the great mushroom clouds of caramel-scented mist that emanated from his open mouth.

"People think I'm this guy who really despises nature," he said, the lenses of his spectacles glinting in the basement's halogen glow. "But that's not true."

Marlo, who was tinkering with some wired-up circuitry over on the far side of the room, swiveled a quarter turn in his ergonomic chair. "To be fair, mate," he said, "you do kind of come across as a guy who despises nature."

"I don't despise it," said Tim, snickering indulgently. "I just point out its limitations. People want to stay being the monkeys they are. They don't like to acknowledge that their brains aren't giving them the full picture, aren't allowing them to make rational choices. They think they're in control, but they're not."

Tim knew what it meant to not be in control. He knew what it meant to experience himself as a desiring-machine, a conductor in a circuitry of need and satisfaction. When he left high school he joined the army. This was before 9/11, and the great boom in the American war industry that followed, and so he was never deployed overseas. When he left the military, he started drinking heavily, and throughout his twenties he was a mess, he told me, a helpless subject of implacable systems within and without. He would awake in the morning and tell himself he wasn't going to drink that day—and he would mean it, he said, but the craving would assert itself, the physical spasms of desire, and he would have little choice but to follow the orders of the chemicals in his brain. The act of drinking was never the result so much of a decision as of a yielding to a force much greater than his own resolve, and he would never know what part of the equation was really him: the need or the resistance; the voice in his head insisting that there could be no drinking today, or the convulsions in his body insisting that there must.

He was a bad drunk: volatile, mean, obliviously driven by rage and self-contempt. His teenage years in the Pittsburgh punk scene, his years as a soldier: all this contributed to what he described as his flair for fistfighting. It was, he acknowledged, a huge character flaw, but it felt glorious to defeat another guy with his bare hands. Even now, he said, every fight he didn't get into was one he had to remember, another one he had to regret.

One day, he told me, he woke up in the hospital, and learned that he had tried to kill himself, and he had no recollection of what had happened. He literally did not know what he was doing. Somewhere along the way, he'd become the father of two young kids, and his relationship with their mother had turned sour and cold and venomous. He was not in control of himself.

After the suicide attempt, after the hospital, he went into AA, and gave up entirely the illusion of free will. He was an atheist, but he relinquished all that he was, as required, to the Higher Power, in which nebulous entity he forced himself to believe, though he never did believe in that belief. And this system worked, this mechanism: he hadn't had a drink in seven years.

When Tim talked about the body, when he spoke of humans as monkeys, or as deterministic mechanisms, he spoke in general terms, but it also seemed clear to me that he was speaking specifically about himself, about his addiction and his overcoming of it. Not being an addict anymore was the start of a journey that would end, somewhere down the line, in not being a human anymore, in no longer being subject to those animal urges and frailties.

In January of 2011, he came across a talk on the Internet by a young Englishwoman who called herself Lepht Anonym. The talk was called "Cybernetics for the Masses," and she spoke about her DIY experiments in extending her own senses, inserting magnets and other devices beneath her skin. Unable to secure the assistance of an actual medical professional, she'd performed these procedures on herself in her own home, sterilizing her equipment–vegetable peelers, scalpels, needles–with vodka. She'd consulted some introductory anatomy textbooks to make sure she didn't damage any major nerves or blood vessels, and gone to work on hacking her body, on becoming a machine.

"Lepht was kind of crazy," said Tim, "but she was also fucking badass. I really admired her."

"She was a hard-core hacker," agreed Marlo.

"To the fucking bone," said Tim. "So when I saw that shit, it was like, Fuck, man, the revolution's started without me."

On an online forum called biohack.me, Tim got to know a Pittsburgh-based engineer named Shawn Sarver, and together they decided to start designing and building their own cyborg technologies. Shawn was also ex-military; he'd enlisted in the air force after 9/11, and had done three tours of duty in Iraq between 2003 and 2005, specializing, as an avionic technician, in the recovery of materials from shot-down planes. To look at the guy, you wouldn't necessarily peg him as a former serviceman: the day I met him, in Tim's basement, he was wearing a tweed sport jacket with velvet elbow patches, his extravagant blond mustache waxed at either end into an opulent coil, and he looked like nothing so much as a villainous dandy from a Victorian children's book. When he wasn't working on the cyborg future, Shawn was a barber in Pittsburgh. (For years now, he had been working his way through what he called "an ancient bucket list of occupations," on which he had so far ticked off such diverse fields as the military, firefighting, electrical engineering, and gentleman's grooming.)

Tim and Shawn talked about how basic army training involves a violent obliteration of your idea of yourself as an individual, and of how, in a different way, a breaking down of the personality, a letting go of the self, is required of you in AA. I suggested that all of this seemed to inform Tim's view of what it meant or did not mean to be a person, his radical and literal remaking of himself. He saw what I meant, but despite his rhetoric about humans as predictable and deterministic mechanisms seemed reluctant to accept such a deterministic interpretation of his own life.

It was Friday night, and the group had just finished its weekly meeting, where everyone—including team members across the country and one or two living abroad—had sat around in Tim's living room and talked about what they'd been working on. A few of us were standing out on Tim's back porch, enjoying the view of the abandoned motel, drinking bottles of some kind of horrible berry-infused beer.

The conversation turned to a guy called Ben Engel, a young grinder from Utah whom some of the Grindhouse people had met recently at a grinder festival in Bakersfield, California. He had built a Bluetooth-enabled gadget that would conduct soundwaves to his inner ear through the bones of his skull. The device would be switched on using a magnet implanted in his finger, and it would, in theory, translate data downloaded from the Internet into compressed audio waves, which he would train himself to interpret using a technique called sensory substitution. He'd been in touch with Grindhouse about his plan, and they had been trying to dissuade him from going through with it, because of the likelihood that it would kill him.

"He just basically Frankensteined this thing together using a bunch of shit," said Justin Worst, a Grindhouse engineer to whom he'd shown the device. "An electric toothbrush charger, a couple cell phone parts. The thing is fucking huge."

Right now, they were trying to convince him to abandon his skull-conduction device and use Grindhouse's technology instead.

"We're just really nervous that the implant would leak through to the brain. It would not be good for the movement to have this dude kill himself," said Tim, who then wandered back into the kitchen and descended into the gloom of the basement. One of his dogs, a terrier named Johnny, trotted out onto the porch and began to take a polite interest in my lower legs. I noted that the dog was himself missing a leg.

"What happened to Johnny's leg?" I asked.

"He got hit by a car," said Olivia Webb, head of safety testing. It was Olivia's last night with the company; after several years in Pittsburgh, she was about to leave for a new job in Seattle.

"And then he ate his leg," said Justin. "Tim and Danielle woke up one morning, and he'd been gnawing on it all night, and they had to go get it amputated."

Working his way thoughtfully through a bag of potato chips, Ryan asked whether, once humans began to successfully merge themselves with machines, we might also extend the courtesy to our pets.

"Would that be ethical?" he said. "Or should we just let them live out their miserable biological lives and then die?"

"We already do things without their consent anyway," said Olivia. "No dog is ever going to be like 'Please take my balls off,' but we do it anyways, for the betterment of them. I mean, Johnny's doing okay with his three legs, but what if he had like a bionic fourth leg?"

Johnny efficiently consumed the few remaining shards of potato chips that had fallen from Ryan's bag, then removed himself to the kitchen at an arrhythmical gallop, as though in passive-aggressive reproach of any such suggestion.

The more time I spent with Grindhouse, the more it became apparent that their ultimate interest was not the augmentation of the human body per se. Which is to say that they were not especially interested in the marginal–and, I would say, highly debatable–degree to which your human life might be made more convenient if you had, say, a subdermal implant that lit up or vibrated when you were facing toward magnetic north. Certainly, they were frustrated by the limitations of the body, and wanted to ameliorate those limitations through technology. Tim, for instance, said that when he first got electromagnets implanted in his fingertips allowing him to sense magnetic fields, he did not suddenly feel exhilarated by his newly expanded sensory capabilities, as most people had assumed he would.

"What I felt," he said, "was terrified. I was like, these things are fucking *everywhere*, and we can't see shit. We are totally fucking blind."

"Exactly," said Marlo. "We can't even see X-rays. I mean, how lame is that?"

But what they were interested in, fundamentally, was something far stranger, far less identifiable, in every sense of the term, than any mere augmentation of our human capabilities. What they were interested in was a final liberation I found it difficult to see as anything other than annihilation.

I was sitting in the basement with Tim and Marlo and Justin while they worked on the new Northstar implant. Wu-Tang's "Protect Ya Neck" was blasting tinnily from a set of desktop speakers, and Tim was nodding emphatically along to the beat while bashing out some code on a laptop. I was perched uncomfortably on the saddle-shaped seat of a workout bench. To no one in particular, to the room in general, I said: "So what's the endgame here? What are you guys looking to achieve, long-term?"

Marlo rotated toward me, a soldering iron poised delicately in his hand, and said that, to speak in a purely personal capacity, he himself wanted to consume the entire universe. Personally, he said, he wanted to become a being of such unimaginably vast power and knowledge that there was literally nothing outside of him, nothing beyond him, that all of existence, all of space and time, was consubstantial with the being formerly known as Marlo Webber.

I said that he would be well advised not to put this on his application form for a U.S. work visa; he laughed, but in such a way that it did not seem that he had previously been thinking of it as a laughing matter. It is possible that he *was* joking. He did, as I have said, perpetually bear the expression of someone who was sharing some private joke with himself. And there was an appealing absurdity, I found, in the extent to which this personal ambition to assimilate within himself the entire universe seemed at odds with the leisurely precision of his Australian drawl, his affable if obscurely aloof vibe. But it didn't seem to me that he was being insincere.

"I'm not sure whether you're fucking with me," I said.

"I'm not fucking with you in the slightest, mate," said Marlo.

"He's not fucking with you," Tim assured me.

"So what's your endgame," I asked Tim. "Are you looking to consume the entire universe, too?"

"For me," he said, "the endgame is when the entire population of humanity, minus a few douchebags, basically flies into space. My goal, personally, is to peacefully and passionately explore the universe for all eternity. And I'm sure as shit not gonna be doing that in this body."

"But what would you be?" I asked. "And would it be you?"

Tim said that what he imagined himself being was an interconnected system of information-seeking nodes, traveling in ever-widening arcs throughout the universe, sharing intelligence across the vastness, learning, experiencing, collating. And his guess, he said, was that this unimaginably expansive system would be as much him as the six-foot-high assemblage of bone and tissue that happened to be his current form.

I was going to say that all of this sounded hugely expensive; I was going to ask who was supposed to pay for it all. But I thought better of it, in the way that you might think better of making a joke about the central tenets of a person's faith after they had taken the trouble to explain them to you.

And we were, it seemed clear to me, encroaching here on the traditional territory of religious belief—a crossover that I felt characterized many of the conversations I had with Tim, and with other transhumanists.

On my last afternoon there, Tim and I were lounging on the

L-shaped arrangement of couches in his living room, talking about the future, when Johnny hopped up beside me and climbed into my lap and began licking me in a frantic transport of unprovoked affection. I felt the animal's clammy breath on my face and in my mouth, the slick heat of its tongue on my nose, and I strove to appear more pleased by the attention than I was.

And we started talking, then, about the question of whether we were our bodies, and whether Johnny therefore somehow existed less, was diminished in some way other than the merely physical, because he had lost part of his body.

I was not quite sure what I believed, but I said that I felt that embodiment was an irreducible and unquantifiable element of existence, that we were human, and the dog was the dog, only insofar as we were these bodies of ours. I talked about my son, and how my love for him was largely, even fundamentally, a bodily experience, a mammalian phenomenon. When I held him in my arms, I said, I felt his smallness, his compactness, the slender bones of his little shoulders, and I experienced the softness and delicateness of his neck as a physical sensation, a tender swelling, a quickening in the machinery of the heart. I often found myself marveling at how little space he took up in the world, how his chest was no wider than the span of my hand, how he was literally a small object, an arrangement of fragile bones and soft flesh and warm unknowable life. And it was this that constituted my love, my animal anxiety and affection for him, little beast that he was.

I asked Tim about his own children, his love for whom he had spoken of to me on a couple of occasions over the last few days, and he said again how he lived for them, how their appearance in his life had saved him from himself. And he agreed that, yes, he had those feelings too, those animal affections and fears.

"What do your kids think of you wanting to become a machine?" I asked. "What do they think of the implants?"

Tim's face was a blank mask of intense focus, as he refilled his vapebox with a homebrew juice given to him the previous evening by a new intern. I wondered whether he had not heard the question, or was perhaps planning to ignore it. I looked at his pale and thin arms, and tried to read the arcane history of his skin: the tattoos, the implant wounds, the coronation of disfigurements.

"My kids understand what I'm doing," he finally said, still insistently focused on the vapebox. "They're totally saturated in it. My daughter, she's eleven. A little while ago, she said to me, 'Dad, I don't care if you become a robot, but you have to keep your face. I don't want you to replace your face.' Personally, I don't have any sentimental attachment to my face, any more than I have a sentimental attachment to any other part of my body. I could look like the Mars Rover for all I give a shit. But she's pretty attached to my face, I guess."

He took a long hit of the juice, and exhaled heavily; a billowing plume of pure whiteness obscured momentarily this face of his to which he had no sentimental attachment, the dark and faintly Asiatic eyes, the fanatically flared nostrils of a proud and strangely driven man.

He said how great Danielle was with the kids. How she was like a mother to them. He said how she'd wanted children of her own, too, and how hard it had been on her, his refusal to become a father again, his insistence on "not participating in the problem anymore."

And then he expressed a sentiment that struck me as essentially religious, in both meaning and expression. "I'm trapped here," he said, nodding down at his chest, at his legs folded yogically beneath him on the couch. "I'm trapped in this body."

I suggested that this made him sound like a Gnostic heresiarch from the second century AD.

Tim shook his head patiently. "But that's not just a religious idea, man. Ask anyone who's transgender. They'll tell you they're trapped in the wrong body. But me, I'm trapped in the wrong body because I'm trapped in *a* body. *All* bodies are the wrong body."

We were, I felt, nearing the central paradox of transhumanism, the event horizon where Enlightenment rationalism, pushed to its most radical extremes, disappeared into the dark matter of faith. As unfair a double bind as it may have been, the more Tim denied any connection between his own thinking and the mysteries of religion, the more religious he sounded.

But perhaps it wasn't so much that transhumanism was a quasi-religious movement, as that it addressed itself toward the fundamental human contradictions and frustrations that had traditionally been the preserve of faith. To experience oneself as imprisoned within the body, with its frailties and its inexorable finitude—fastened to a dying animal, as Yeats put it—was a fundamental condition of being human. It was, on some level, in the nature of having a body to want out of it.

"Today," wrote D. H. Lawrence, "man gets his sense of the miraculous from science and machinery, radio, airplanes, vast ships, zeppelins, poison gas, artificial silk: these things nourish man's sense of the miraculous as magic did in the past."

And just as the human need for mystery, for cosmic awe, was now increasingly satisfied by science, the longing for some promise of redemption had similarly become the inheritance of technology. Although he would not have put it in these terms, this was ultimately Tim's message, the message of the cyborg: that we would eventually be redeemed of our human nature, of our animal selves, and that all we had to do to secure this redemption was to let technology into our mortal bodies, thereby achieving a communion with machines, a final absolution from ourselves.

Faith

THERE WERE MALFUNCTIONS of equipment; things did not proceed frictionlessly. My journey from San Francisco to Piedmont for the conference on transhumanism and religion was beset on all sides by minor difficulties. From the Mission, where I'd rented an Airbnb place for the few days I was in town, I took the BART across the bay. It was eight-thirty or so on a Saturday morning, in the middle of a ruthless May heatwave, and downtown Oakland was deserted by all but a loose cohort of the afflicted and unhoused. This gave the place an air of sorrowful aftermath, as though in some efficient and bloodless apocalypse all souls had been raptured, but for those tainted by poverty.

I needed to get to Piedmont by 9 a.m., and there was not a cab to be seen. I had maxed out on data roaming within minutes of landing in SFO two days previously, and so had no way of Ubering or Lyfting myself the remaining five miles east to the conference venue. I felt divested, as though bereft of some irreducibly human faculty. After an interlude of vacillation (Should I look around for a café with WiFi? Should I get some quarters and find a pay phone? Were pay phones even a thing anymore?), I eventually flagged down a cab in the manual style, the whole business of which had already begun to feel like some whimsically old-timey affectation. On arrival in Piedmont, there were further complications. My driver, whose English was only barely functional, was relying on a dash-mounted-smartphone-and-Google-Maps system to get us where we needed to go, and Google Maps was doggedly refusing to acknowledge that where we needed to go even existed. By the time the driver eventually pulled up outside the veterans hall, it was a good fifteen minutes past the conference's start time of nine, and I was aware that I may already have missed out on some high-grade material.

A group of men were gathered in conversation toward the rear of the hall, one of whom was Hank Pellissier, the conference's organizer. Hank was in his late forties, with short gray hair, but he had the trim stature, and something of the awkward ebullience, of a teenage boy. This impression was enhanced by his buoyantly youthful attire (rainbow-striped T-shirt, bright green pants, Seinfeldian tennis shoes). I wanted to announce my presence, to thank him for setting me up with a press pass, and for connecting me with various people he thought I might be interested in speaking to. He welcomed me in a manner both warmly enthusiastic and slightly absent, and quickly introduced me to the rest of the clustered miscellany of white American manhood.

There was a stocky and bearded and affable young dude from Nashville who was somehow both a transhumanist and a born-again Christian. There was a sixtyish professor of Systematic Theology at Pacific Lutheran Theological Seminary in Berkeley; a burly man in an olive drab field jacket with pockets both plentiful and capacious, both zipped and buttoned, a garment such as you'd likely favor if you felt the Rapture might potentially kick off while you were out of the house. There was a somber transhumanist Buddhist from Las Cruces, New Mexico. There were two Mormon transhumanists from Utah. (I knew, from long months of Internet stalking, that the Mormons were a small but vocal contingent within transhumanism, and that this had to do with some unexpected synergies between the movement and the beliefs of the Church of Jesus Christ of Latter-day Saints.) And there was a guy called Bryce Lynch, a pale and intense and bespectacled cryptographer in his late thirties, with hair that was both receding and long, and a manner that was both chipper and aloof. I asked Bryce whether he belonged to any faith, and he equivocated a moment before telling me that he practiced a modern form of Hermeticism, an esoteric pagan religion that had more or less peaked in late antiquity. I pressed him gently on the matter, but he seemed a little reticent, which is maybe what you'd be wise to expect from a cryptologist who was also a practicing hermeticist.

Bryce wore a black T-shirt emblazoned with a message whose meaning was obscure to me–

I DON'T ALWAYS TEST MY CODE,
BUT WHEN I DO,
I DO IT IN PRODUCTION

–but which I took, almost certainly wrongly, to be some kind of sexual innuendo rooted in a programming pun. One of the Mormon transhumanists was sufficiently amused by the shirt to ask if he could take a photo of it. Bryce obliged, striking a heroically wide-legged stance, chest pushed forward and arms waggishly akimbo; this posing drew yet more attention to the shirt, whose message was now provoking mild merriment among the group as a whole. That I could not share in this merriment, that I could only chuckle politely and hope that I would not be called upon to make comment on its meaning, made me strangely self-conscious about my difference from these people, a difference that came to seem more profoundly irreconcilable the more I considered it. I was basically illiterate in the language of technology, was the thing. I was a *user* of technology, a passive beneficiary of its many advances, while knowing next to nothing about it per se; these people, though, these transhumanists, were rooted in the intimate logic of machines, grounded in the source code of our culture.

A tall, silver-haired man in a dark suit appeared in the doorway, and Hank excused himself to go and speak to him. The professor of Systematic Theology and the Buddhist transhumanist glanced at one another knowingly, as if to acknowledge the falling into place of some crucial element. They spoke among themselves for a moment, and I took out my phone, wondering whether it would be bad form to just start recording conversations at this point on my phone's voice recorder app. Hank was now helping the silver-haired gentleman to set himself up at a trestle table at the very back of the room, some distance behind the last row of seats.

The professor of Systematic Theology canted himself toward me, sidemouthed a name: "Wesley J. Smith."

I nodded, more knowingly than I had any right to be nodding, given that I'd never heard of the man.

Smith, Mike informed me, was a regular contributor to *National Review*. He was a religious man, a convert to the Eastern Orthodox faith, who had in recent years been carving out a niche for himself as a conservative commentator on questions of bioethics, which is how he'd wound up writing on transhumanism. He was here to report on the conference for an interfaith journal called *First Things*.

In 2013, he'd published a piece there called "The Materialists' Rapture," in which he'd criticized transhumanism on the grounds that it was essentially a religion, an odd point of attack from a guy who presumably sees religion as basically a good thing. "Proselytizers for 'transhumanism,'" he writes, "claim that through the wonders of technology you or your children will live forever. Not only that, but within decades you will be able to transform your body and consciousness into an infinite variety of designs and purposes, with self-directed evolution leading to the development of 'post human species' possessing comic book character-like super abilities. Indeed, one day we will be god-like." He points out the parallels between transhumanism and Christianity in a way that seems to me to be perfectly accurate, drawing a particular comparison between the Rapture of Christian eschatology and the concept of the Singularity. Both are projected to occur at a specific time; both will ultimately lead to the final defeat of death;

both will usher in an Edenic age of harmony in a "New Jerusalem"—respectively, in heaven and here on earth; both Christians and Singularitarian transhumanists expect to be furnished with brand-new "glorified" bodies, and so on.

I didn't see much to take issue with in any of this, aside from the implication that these links with religion somehow discredited transhumanism. It seemed to me that transhumanism was an expression of the profound human longing to transcend the confusion and desire and impotence and sickness of the body, cowering in the darkening shadow of its own decay. This longing had historically been the domain of religion, and was now the increasingly fertile terrain of technology. Wesley J. Smith saw transhumanism as an abomination, a perversion, a shallow and grotesque parody of religion. I saw it as a new expression of these same immemorial yearnings and frustrations.

Smith was established now with his laptop and trestle table at the rear of the room, a setup that gave off the impression of a kind of journalistic rampart, behind which he had cordoned himself off from the people and ideas he was here to cover, to *report* on. I was fascinated by the frank and easygoing hostility of this gesture, the bluntness with which it announced Smith's detachment from the proceedings.

Back at the apartment that night, after I'd spent ten or fifteen minutes looking anxiously through my notes, wondering what if anything they amounted to, I checked my email and saw, via the Google News Alert I had set up for the word "transhumanism," that Smith had already published a piece on the conference, a blogpost for *National Review* that had been posted at 7:33 p.m., while he was still sitting at his desk at the back of the conference hall. It was no masterpiece, certainly, and did little to advance or develop the position he'd already taken on transhumanism. "For now," he wrote, "I have to say that my previous opinion of transhumanism as a materialistic religion—or perhaps better stated, a worldview that seeks to obtain the benefits of religion without submitting to concepts of sin or the humility of believing in a Higher Being—are being substantially borne out." (From a career perspective, I thought, perhaps there was something to be said for knowing what you thought about things, for seeing the world in such a way as your opinions were always being substantially borne out.)

Hank's introduction to the event was a characteristically strange and meandering oration. (His background was in the Bay Area punk scene, where he had made a name in the 1990s as a performance poet with the pseudonym Hank Hyena. His work, he'd told me, tended toward a kind of erotic absurdism.) Hank talked at some length about his own complex history of religious association, something in which he seemed to take a counterintuitive pride. A few years back, he'd taken his family to live in a Quaker community. He soon "got waylaid by doubts" and briefly became a militant atheist, and then got turned off by the militance of militant atheism, and became a transhumanist. This happened more or less by chance, he said, when an editor he'd worked with in the past got a new job with a transhumanist publication called *H+* and asked him to do some writing. He agreed, and even though he'd never heard of transhumanism, he got involved in the movement almost immediately, realizing, as he put it, that he had always instinctively been a transhumanist but just never knew it was an established thing. He was now, he said, in the middle of an on-and-off flirtation with Judaism.

"I'm a sperm donor to a lesbian couple," he explained, "one of whom is a rabbi. I may convert for the sake of my biological son. I'm still thinking it over."

I went on to hear a great many strange things spoken that day, by a great variety of people.

I heard a sex shop proprietor who'd written a guide to strap-on penetration called *Bend Over Boyfriend* speak of her spiritual development as a Wiccan.

I heard one of the Mormon transhumanists speak of how he was a transhumanist because of and not despite his Mormonism.

I heard a man who ran an independent publishing house dedicated to topics of extremely niche esoteric interest speak at considerable length about *The Urantia Book*, a gigantic work of cosmogony, supposedly dictated to its authors by the ancient aliens who created the human race; I heard him speak of the rebellion of Lucifer, of his own personal conviction that the Nephilim are the source of the Illuminati bloodlines, and I thought how strange it was to hear these things being spoken by a man in boat shoes, comfortable jeans, and a smart blue sport jacket; I heard him speak, finally, of his part in an expedition to find the sunken land that was the location of both Atlantis and the Garden of Eden, but he was cut off by Hank on the grounds of having already spoken for far too long, and so I never heard him speak of whether he had in fact discovered this sunken land.

I heard the Buddhist transhumanist, whose name was Mike LaTorra, speak of his belief that he was in fact already living eternally through reincarnation, and of how he simply wished to do so in a body superior to the one he currently possessed.

I heard a Seventh Day Adventist pastor with the pleasingly literary name Robert Walden Kurtz, who knew something of cults (having been personally acquainted with a guy who died at Waco after joining David Koresh's Branch Davidian sect in the 1990s), speak of how transhumanism could easily lend itself to such extreme and eccentric spiritual offshoots.

I heard a man named Felix Clairvoyant, a certified massage therapist with a PhD from the Institute for the Advanced Study of Human Sexuality—who wore a semitranslucent crepe shirt and black slip-on shoes with no socks—speak of his belief as a Raëlian that the human race was the creation of scientists who came here in UFOs thousands of years ago.

And I could not but be impressed by the richly ecumenical vibe of all of this, by the extent to which these people wished to learn about the beliefs of the others, incompatible though they may have been with their own. The Mormon transhumanists seemed surprisingly well informed about the practices and beliefs of Wiccans; the

Adventists were keen to engage the Buddhist in friendly and sophisticated colloquy; even the Raëlian massage therapist was engaged in a spirit of respectful curiosity by the Atlantis expedition guy in the sport jacket and boat shoes.

These things took many hours to hear, and to hear them I was obliged to sit in a steel-framed chair, much like the sort of chair I sat in as a schoolboy, and which was now causing my lower back to hurt, and my buttocks to become somehow both painful and numb, and my legs to become stiff, and my thoughts to turn to the inevitable decline of the body–to mortality itself, and to the many other disadvantages of being fastened to a dying animal.

During an afternoon break in the conference, I walked with Mike LaTorra to a sandwich shop around the corner from the veterans hall. We sat outside in the warm California afternoon, and he spoke about the various ways in which his Buddhism and his transhumanism complemented and contradicted one another. In a rich and soothing baritone, with an air of oceanic and slightly sad tranquillity, he explained that Buddhism, and the practice of meditation specifically, were oriented toward the relief of pain, toward the attainment of a plane of consciousness beyond and above the striving and anxiety and misery of normal human experience.

"Life is suffering," said Mike, calmly and methodically working his way through a small bag of organic beetroot chips. "If you survey any group of people at any point of history, the vast majority would tell you that, yeah, things could definitely be better. We're not in hell, but you might think of the world as being one elevator floor up from hell."

The Buddha's message, said Mike, was in some sense a transhumanist one: life is suffering, yes, but there is a path that leads to the end of suffering. He saw Buddhism and transhumanism, in this sense, as differing approaches to the overall problem of life being basically unsatisfactory. He spoke about the esoteric idea of spiritual ascent within Buddhism, the four stages that a person passes through on the way toward full enlightenment. This notion of attaining a higher plane of personhood, he said, was one that he found to be deeply compatible with the transhumanist ideal of transcending the condition of humanity through technology.

I was curious about the extent to which the transhumanist belief that the mind could exist separately from the body was at odds with the Buddhist idea of embodied existence—that the self is not some rarefied entity distinct from the animal in which it is contained.

"Well, there's more than one school of thought on that within Buddhism," he said. "In Zen Buddhism, which is what you're thinking of, there is no separation between me and my body. There is no ghost in the machine. But in Theravada Buddhism, which is the oldest form of the tradition, we are not the body; the body is something to be rejected and held in contempt. To be transcended."

He spoke about affirmations that the newly ordained Theravada monks were made to recite, in which the body was rejected as a site of corruption and decay. "There's a kind of revulsion there," he said, "in those early Buddhist texts. A revulsion against the human body, against biology."

They fucked us up, our first mum and dad. That decision to eat from the tree of knowledge, to heed the serpent's counsel that doing so would make them as gods: that was the moment when everything got shot to hell. As far as the Judeo-Christian tradition is concerned, the whole human condition is a punishment for an audacious infringement, way back in those early days: that first disruption of the knowledge economy.

And it could all have been so different. In the seventeenth century, in the first blush of Enlightenment's dawn, Adam was a kind of proto-transhumanist ideal. According to the philosopher and clergyman Joseph Glanvill, the first man was blessed, among other things, with superhuman sight: He "needed no Spectacles. The acuteness of his natural Opticks... shew'd him much of Coelestial magnificence and bravery without a Gallileo's tube." The occultist and herbalist Simon Forman claimed that the forbidden fruit had introduced a lethal toxicity into the bodies of our first parents, causing a degeneration that worsened through the ages. Adam, he wrote, "becam monstrous and lost his first form and shape divine and heavenly and becam earthy full of sores and sickness for evermore." The apothecary Sir Robert Talbor wrote that the soul and body of man "have deviated from the first perfection," and that "the Memory is subject to fail, the Judgement given to erre, and the Will often known to rebel, and become a voluntary slave to passion; so is his Body subject to so many infirmities."

In *The Advancement of Learning*, Francis Bacon, who is often seen as the founder of the modern scientific method, addressed the ancient shame that haunted the idea of knowledge in the Judeo-Christian imagination, the original unity of learning and sin. He writes of hearing learned men "say that knowledge is of those things that which are to be accepted of with great limitation and caution; that the aspiring to overmuch knowledge was the original temptation and sin whereupon ensued the fall of man; that knowledge hath in it somewhat of the serpent, and, therefore, where it entereth into a man it makes him swell."

But Bacon believed that we could reclaim something like our prelapsarian perfection—our original state of immortality and divine wisdom and peace—through the application of science. The way back to Eden, in other words, could only be found by continuing along the path of our first divergence. Toward the end of his life, he was given to wondering about the possibility of reversing, through science, the consequences of original sin. The prolongation of life was one of the foundational aims of Bacon's "Great Instauration," his proposed reformation of scientific knowledge, itself modeled after the divine work of the six days of creation. Despite Bacon's apparent millenarian belief that the earth was approaching its final age, writes the cultural historian David Boyd Haycock, he "rejected pessimistic views of natural history. If this *was* the earth's dotage, for Bacon it was to be a mature old age of profound wisdom and learning, in which European scholars would pluck the final fruits of God's benevolent creation. Natural philosophers would take full advantage of all that had gone before them, restoring the greatness that had once been Adam's. Only then, when this last great age of progress had been fulfilled, would the world be fit for Doomsday."

Bacon died in his mid-sixties, by no means a bad run in those days. But he died a death whose cheap irony was beneath him: according to his contemporary John Aubrey, he caught pneumonia while burying with his bare hands a freshly slaughtered chicken in the snow, thereby proving that freezing could preserve the flesh of animals and men.

We are always trying to get back to that state of wholeness that preceded the Fall, the split, the loss. It is knowledge that we feel will return us to innocence. As the eighteenth-century German writer Heinrich von Kleist puts it in his strange and brilliant essay "On the Marionette Theatre":

> We see in the organic world, as thought grows dimmer and weaker, grace emerges more brilliantly and decisively. But just as a section drawn through two lines suddenly reappears on the other side after passing through infinity, or as the image in a concave mirror turns up again right in front of us after dwindling into the distance, so grace itself returns when knowledge has as it were gone through an infinity. Grace appears most purely in that human form which either has no consciousness or an infinite consciousness. That is, in the puppet or in the god.... But that's the final chapter in the history of the world.

The conference's last panel had just ended. I was gathering my effects, wondering how I was going to get back across the bay to San Francisco, when Hank came by to let me know about something that was about to happen, something he thought I might be interested in. Jason Xu, the Silicon Valley community organizer for Terasem, was setting up for a little gathering in a room off the main hall. I had read about Terasem, and it seemed to be the closest that transhumanism had come to generating a genuine religious offshoot. It was a faith, or "movement," based in the idea of "personal cyberconsciousness," in the spiritual dimension of things like mind uploading and radical life extension. I had read about Jason Xu, too—about a protest he'd helped to organize recently, the first-ever transhumanist street action in the U.S. Outside Google's headquarters in Mountain View, he and a small group of fellow transhumanists had stood with placards reading "IMMORTALITY NOW" and "GOOGLE, PLEASE SOLVE DEATH." The idea of the protest seemed counterintuitive, given that solving the notoriously awkward problem of death was precisely what Google, having just pumped hundreds of millions into its biotech research and development group Calico, was now setting out to do. (In this sense, it wasn't so much a protest as an organized encouragement that Google keep up the good work; either way, they were still ejected from the premises by security.)

I hadn't known that Jason was going to be holding an actual meeting at the conference, and so I was excited to get to sit in on it. I would, however, be lying if I told you that the offer of free pizza had not had its own attractions. And I was not alone in this motivation. None of us in this little group that had assembled for the Terasem meeting—me, Mike LaTorra, Bryce Lynch, some guy called Tom—had yet been liberated from the brute imperatives of our animal bodies. And so each of us was, for now, bowed in silent communion with his own slice of pepperoni pizza.

Jason suggested that we all say something about ourselves and why we had come here. He pointed to Tom, and asked if he'd like to start us off, and Tom gave it a go, but it quickly became apparent that he had too much pizza in his mouth to really pull off a coherent self-introduction, so Jason then nodded at Bryce, who was sitting next to Tom, but Bryce shook his head and gestured face-ward to indicate that his own mouth was also too committed to pizza to really convey any kind of useful autobiographical particulars at this juncture, and so then Jason looked at his watch and acknowledged that we should probably just wait until we were all finished chewing before we started the meeting proper. In this time, he handed each of us a bound booklet of photocopied pages entitled *THE TRUTHS OF TERASEM: A Transreligion for Technological Times.*

Once all five of us had briefly introduced ourselves, Jason said a few prefatory words. "Transreligion," he explained, meant that you could join the church even if you were already a practicing member of some other religion. Insofar as Terasem is a religion at all, it seems closer to Buddhism than any of the Abrahamic faiths—at least in the narrow sense that there is no deity at its center, no one entity calling the celestial shots, demanding the fealty of prayer and obedience. The first truth of Terasem, according to the booklet, is that "Terasem is a *collective* consciousness dedicated to diversity, unity and joyful immortality."

Probably the most attention-grabbing aspect of the religion was one that Jason totally neglected to mention here, but which I knew about from research I'd done online: the practice of "mind-filing," an idea taken from Kurzweil's *The Singularity Is Near.* This is a daily techno-spiritual observance, whereby you upload some measure of data about yourself—a video, a memory, an impression, a photograph—to one of Terasem's cloud servers, where it will be stored until such time as an unspecified future technology will be capable of reconstructing, from this accumulated data, a version of you, of your very soul, which can in turn be uploaded to an artificial body, that you might live eternally, blissfully, unencumbered by your mortal flesh. It's not totally clear whether the practice is meant to be symbolic; the whole thing was a little sketchy in terms of details.

From a shoulder bag beneath his seat, Jason withdrew a MacBook Air. Propping it open on his knees, he queued up a recording of

Terasem's anthem, "Earthseed." It began with a minor-key arpeggiating piano, into which stately setting a woman's soulful vibrato interposed itself. The sound quality and volume from the laptop speakers were poor, and whatever emotions this anthem was intended to stir remained unstirred, at least within my own doubtful heart, but the words could nonetheless be made out with clarity:

> Earthseed, come to me!
> Earthseed, come to you!
> Earthseed, one are we!
> Earthseed, that's the truth!
>
> Truth for you, truth for me!
>
> Earthseed, stand with us!
> Earthseed, march with us!
> Earthseed, strengthen us!
> Earthseed, consciousness!
> Collective... consciousness!

The song, Jason explained, had been composed by Terasem's founder, Martine Rothblatt, who was also responsible for the piano accompaniment and flute solo. Martine was a particularly strange and intriguing figure, even for a transhumanist. She had made her considerable fortune from founding the first-ever satellite radio company, Sirius FM, and had later established the biotechnology company United Therapeutics, of which Kurzweil is a board member. I'd read a *New York Times* article about Bina48, the talking robot doppelgänger she had made of her wife, Bina, with whom she had had four children before undergoing a sex change in 1994, having spent her first forty years as Martin Rothblatt. Over the last decade or so, she had spearheaded a campaign to secure peace in the Middle East by making Israel and Palestine the fifty-first and fifty-second states of the U.S. In all of this, she seemed a parodic extrapolation of the figure of the billionaire individualist.

Her advocacy of transhumanism was intimately linked with her status as a transgender woman; in what I had read of her writing, the rhetoric of liberation was always in play–liberation not just from gender, but from the fact of embodiment, the flesh itself. ("It is the mind that is salient, not the matter that surrounds it," as she put it in an essay called "Mind Is Deeper than Matter: Transgenderism, Transhumanism, and the Freedom of Form.")

Jason announced that we would be reading aloud this evening from section three of *The Truths of Terasem,* taking the short subsections in counterclockwise turns around our little group. There was a flurry of page flipping, a brief salvo of throat clearing, and Jason began.

"*Where is Terasem?*" he read. His voice was flat and expressionless, and he kept his eyes on the page as he spoke. "Terasem is *everywhere* and everywhen consciousness organizes itself to create diversity, unity and joyful immortality."

Jason nodded at Mike, who sat to his right.

"*Everywhere* means physical and cyberspace," read Mike in his rich and equable baritone, "real and virtual reality, because vitology can thrive in many spaces."

Jason nodded toward me. I read:

"*Spaces* where Terasem thrives are limited only by their ability to support consciousness." I pushed hard against the words, enunciating them louder and more clearly than was strictly necessary; hearing them spoken aloud in my own voice seemed to heighten their absurdity. (I was reminded of the morning assemblies I attended thrice-weekly throughout secondary school, in which my fellow students and I were obliged to read Bible passages and sing hymns, and I recalled the strangeness of those words in my mouth, the invocations and supplications to a god I could never imagine as anything other than an unreal abstraction, a void around which the world had been arranged.)

The baton was then passed to Tom, who it turned out had a very serious speech impediment. The room went into that strange state of almost meditative suspension that prevails in the presence of extended stuttering. He was about halfway through his sentence—"Physical places that support Terasem consciousness include the earth, heavenly bodies and colonies in space"—when Jason leaned forward and flatly informed him that it was totally okay to skip syllables, putting it in such a way as to imply this was obviously the sensible option. I wondered whether Jason was really cut out for this whole community outreach venture, even if the community he was reaching out to was Silicon Valley.

There now entered, noisily, a latecomer, a sort of classically hippieish guy who looked to be in his late sixties. His hair was long and entirely gray, and his beard, likewise long and gray, was forked into two tapering dreadlocks. He sat down next to me, and looked around at the group with an air of open amusement. He was a ghost of California past, this man, come to haunt its present, its future.

"I'm totally new here," he drawled, superfluously. "What am I supposed to do?"

Jason told him to tell us all a little bit about himself. It was difficult to say whether he was mildly irritated, or whether that was just his social default mode.

"What do you want to know?" said the man, whose name I didn't catch, or who never offered it to begin with.

"Well, how did you find out about the conference, for instance?"

"I don't know, man," he said, with a slow and needlessly elaborate shrug. He seemed quietly amused by himself, or by the situation, or both. "Just surfin' the Web, I guess."

We returned to our readings.

"Instantiating yourself into software form is like getting an education—some things change and some things don't," read Mike.

"Never fear multiple versions of yourself—they'll all update each other just like family does," read Bryce.

"Creating your cyber-self accelerates your joyful immorality," read the bearded latecomer.

Jason interjected. "That's supposed to be 'joyful immor*tality*,' actually."

"Says 'immorality' here."

"No, it should definitely say 'immor*tality*.'"

"Yeah, well, it doesn't. There's no 't.'"

"I don't see how–"

The man brought his copy of the handbook closer to his face, the better to read it.

"Oh yeah, sorry, my bad," he said, in a manner that did not convey much in the way of remorse. "There's the 't.'"

We continued reading aloud for a further five minutes or so, each of us in turn announcing things that he likely neither believed nor understood: that we should never say goodbye to deceased loved ones because we will see them again in cyberspace; that living in an emulated environment beats living "raw" because suffering will be "deleted"; and so on. The more we read, the less sense I could make of any of it. It was an impenetrable torrent of words now, an overwhelming profusion of mere assertion. *Effective immortality is achieved by dispersing throughout the galaxy and universe encoded data emulations of reality. Nature is honored by recreation of the past and immortal preservation of joy and happiness.*

Eventually Jason announced that the evening's reading had concluded, and he asked whether anyone had any questions. Given that I had already outed myself as a writer, I was aware of some vague professional and social obligation to ask him something about the movement, but nothing presented itself to me. I was still feeling a little overwhelmed by the unchecked deluge of proclamations that had just concluded.

"No questions, then?" said Jason.

The old hippie raised his hand with a facetious hesitance.

"I got a question," he said. "Can I get a slice of that pizza?"

There was a general silence as he made his way over to the wheeled hospital-style tray on which the pizza was laid out, as he sat back down with his slice of pepperoni and cheese. As he ate, he flipped backward through the pages of the handbook and then held it outward, hefting it in his palm. In a voice muffled by the pizza he had yet to finish chewing, he asked Jason why it was that the handbook didn't have any Web address on it.

"Like, if I get home and I want to look up this thing, this whole Terasem deal? I won't know how to find the website."

"You can just Google 'Terasem,' I guess?" said Jason, who was no longer trying to hide his irritation with the attitude of this puckish interloper toward his meeting, his movement, his faith.

"Yeah, okay, sure. But just in terms of like PR or whatever, it seems like it'd be a good thing to have on there. The Web address. Just for the sake of people's convenience, is all."

Jason then explained that we would not actually be taking the handbooks home, that he would in fact be getting them back from us when we were finished here, which (he looked at his watch) would have to be pretty soon.

At this point, I panicked a little. Up until the early afternoon, I had been relying quite heavily on my phone as a mnemonic prosthesis, for retaining things I would need to draw on later, and which I couldn't trust my own defective memory to hold on to—photographs of people whose appearance I wanted to recall, audio snippets of conversation, the occasional brief video clip. Pretty quickly, my phone had run out of memory, and, because I had maxed out on data roaming and therefore could not access cloud storage, the only way I would have been able to continue recording things on it would have been to ruthlessly delete photographs and videos of my wife and son, which I was not prepared to do.

So I had since then been relying largely on my own haphazard note-taking, scrawling impressions and quotes on whatever came to hand. For the last hour or so, I had been jotting down these quotes and impressions in my copy of the Terasem handbook, and so I was reluctant to hand it back to Jason, because I would be needing these notes in order to reconstruct the scene in writing. A more acute cause of this reluctance was that some of the impressions I had been jotting in the handbook were quite blunt in their portrayal of Terasem, and of Jason himself. ("*'It's okay to skip syllables'? Jason = kind of an asshole.*") I had no wish to sour my relationship with a potential source, or to cause myself any undue awkwardness, and so I did the only thing I could think to do at that point, which was to grab my jacket from the back of my chair and make straight for the door, head down, heedless of any questioning gazes that may have followed me from the room.

Outside in the otherwise empty foyer, one of the Mormons was sitting alone, bowed in the whitish glow of his laptop screen. I asked him for the WiFi password, and he gave it to me. I opened the Uber app on my phone, and summoned a car to my location, unknown though it was to myself, and gave thanks for the graceful intercession of technology.

Please Solve Death

IN THE DAYS and weeks that followed the Terasem gathering, I thought frequently of Jason Xu's "protest" at the Google campus. I kept thinking, in particular, of the "GOOGLE, PLEASE SOLVE DEATH" placard. The phrase, for all its absurdity, seemed to enclose within itself the strange cluster of desires and ideologies at the heart of transhumanism, with its faith in the power and benevolence of techno-capitalism.

It was less a protest than a supplication, a prayer. *Deliver us from evil.* Save us from our bodies, our fallen selves. *For Thine is the kingdom, the power and the glory.*

The word "solve," in this context, seemed to me to encapsulate the Silicon Valley ideology whereby all of life could neatly be divided into problems and solutions–solutions that always took the form of some or other application of technology. Whether the problem was having to pick up your own dry cleaning, or negotiate the complexities and uncertainties of sexual relationships, or face the reality that you would one day die, that problem could be hacked. Death, in this view, was no longer a philosophical problem; it was a technical problem. And every technical problem admitted of a technical solution.

I remembered what Ed Boyden had told me in Switzerland: "Our goal is to solve the brain."

In the foreword to a 2013 book on the science of life extension,

Peter Thiel wrote that the key distinction between computer science and biological science, that "computers involve bits and reversible processes" while "biology involves stuff and seemingly irreversible processes," was on the verge of dissolution; computational power, he argued, would be brought increasingly to bear on the domain of biology, permitting us to "reverse all human ailments in the same way that we can fix the bugs of a computer program." "Unlike the world of stuff," he wrote, "in the world of bits the arrow of time can be turned backward. Death will eventually be reduced from a mystery to a solvable problem."

Solve the brain. Solve death. Solve being alive.

Among the life extension researchers who had received funding from Thiel was an English biomedical gerontologist named Aubrey de Grey. De Grey was the director of a nonprofit called SENS (Strategies for Engineered Negligible Senescence). He had attracted considerable notoriety for the claim that he was currently developing treatments which would enable human beings now living to extend their life spans indefinitely. It was his specific contention that aging was a disease, and furthermore a curable one, and that it should be approached as such: that we should be prosecuting a great counteroffensive against our common enemy, mortality itself.

I'd been aware of Aubrey's work for some years before I met him. He was one of the most prominent figures within the transhumanist movement. Max and Natasha Vita-More had both spoken approvingly of his work, as had Randal Koene; he had been the subject of a handful of books and documentaries, and of a profusion of variously credulous and dismissive newspaper articles. Among the ideas he had popularized (through, among other channels, a widely consumed 2005 TED talk) was something referred to as "longevity escape velocity." This was the notion that the pace of technological advancement in the area of life extension would eventually increase to the point that, for every year that passes, average human life expectancy increases by more than a year—at which point, the theory goes, we put a comfortable distance between ourselves and our own mortality. Over the past century or so, life expectancy had been increasing at the rate of about two years per decade, but the optimistic expectation within the life extension movement was that we would soon reach a point where the ratio flipped—thereby, as de Grey put it, "effectively eliminating the relationship between how old you are and how likely you are to die in the next year."

This idea of longevity escape velocity was something like an article of faith among transhumanists and life extension enthusiasts. It was an idea that Max More, for instance, had raised a number of times when I spoke with him—as the source of his hope, for instance, that he would not himself have to rely on the fallback of cryonic suspension to ensure the radical extension of his own life. And it was the central premise, too, of Ray Kurzweil and Terry Grossman's 2004 book *Fantastic Voyage: Live Long Enough to Live Forever,* which argued that if middle-aged men like its authors could simply live to the age of 120, they would then be in a position to never die at all.

I met Aubrey one August morning at a cavernous bar near Union Square in San Francisco, right across the street from the Hilton where he'd just given a talk at a conference of real estate investors. It was shortly after breakfast, and Aubrey was blowing the froth off what may or may not have been the first pint of the day.

Physically, he was an extraordinary proposition: long and somber as a scarecrow, and exhibiting an immensely unreasonable beard, a Rasputinian profusion of wiry russet that terminated chaotically somewhere around his lower rib cage. This beard, for which he was almost as renowned as for his Promethean claims, exerted an almost literally overpowering influence over my interaction with him, not merely in its rich source of visual distraction, but also in the effect it had on his speech, which emerged as somehow both stentorian and muffled, so that for all the dramatic resonance of his disquisition, I occasionally had to ask him to repeat himself.

For the last few years, Aubrey had been dividing his working life evenly between Cambridge and California. He'd flown in from Heathrow late the previous evening, though he betrayed no obvious signs of jet lag, a condition to which he anyway claimed outright immunity. In the last few years, he had moved most of SENS's operations to Silicon Valley, where the culture was a great deal more amenable to his vision of indefinite regeneration and youth, the possibility of a final triumph over death.

"I find," he said, "that there are a higher proportion of people here who are visionaries, who have not lost the ability to aim high."

He dragged a hand downward through his beard, settling to his task. He spoke in the unmistakable drawl of the English upper class, the congenital weary irony.

Although Thiel was one of SENS's major sources of philanthropic donation, by far its largest funder these days was Aubrey himself. On the death of his mother in 2011, he had inherited £11 million worth of property in the London borough of Chelsea, most of which he'd avoided paying any tax on by funneling it into SENS, a registered charity.

But finding a cure for aging was an expensive business. He had an outreach team to pay for, a full-time staff of scientists. By Aubrey's own reckoning, SENS had about another year left of that inheritance money. And so, when I met him, he was focused almost exclusively on increasing external sources of funding, which was why he'd just been hard-selling the prospect of eternity to a roomful of wealthy Bay Area real estate investors across the street, and, in a less direct fashion, why he was talking to me now.

Aubrey was, as it happened, quite gifted in the necessary arts of persuasion; early in our conversation, he caught a whiff of my own skepticism and proceeded ruthlessly, if not entirely effectively, to interrogate and undermine its underlying assumptions.

He first set about arguing me out of any ambivalence about the desirability of eradicating human mortality. People's standard reasons for rejecting the principle of radical life extension–that it would somehow rob us of our humanity, that life was given meaning by its finitude, that living indefinitely would actually be hellish–were "embarrassingly infantile and idiotic" rationalizations. Death, he said, was our captor, our tormentor; and we dealt with this situation through a kind of Stockholm Syndrome. This was beneath contempt.

The brute fact of the matter, he said, was that aging was a human disaster on an unimaginably vast scale. It was a fucking massacre out there, a methodical and comprehensive annihilation of every single person, and he was one of a tiny handful of people taking it seriously for the humanitarian catastrophe that it was.

Such was the rhetoric. Calculated, impassioned, performative.

He said: "For every day that I bring forward the defeat of aging, I'm saving *a hundred thousand fucking lives*!" He brought his fist down hard on the distressed wood of the tabletop.

He said: "That's thirty September 11ths every week! That is thirty World Trade Centers I'm preventing."

The science of regenerative medicine was complicated, but Aubrey had at his disposal an array of simplifications for the lay interlocutor. Among his favored rhetorical gambits was to ask you to think of your body as a classic car, as a complex system of interlocking mechanisms that, through regular maintenance, could be kept more or less indefinitely in a roadworthy state.

"Human bodies are basically just machines," as he put it in a 2010 TEDx talk. The idea, as such, was that we "go in and regularly repair the damage so that we can postpone the time at which the damage is so extensive."

"It's all about restoring the molecular and cellular structure of the body to the state it was earlier in adulthood," he told me now. "What that amounts to, overwhelmingly, is just repairing the various types of damage the body does to itself from the time we're born, as a side effect of basic operation."

He then explained his two-part conception of SENS's project.

"SENS 1.0," in which the organization was now largely engaged, involved various therapies he claimed would be possible to develop within the next two to three decades, given sufficient funding. These therapies, he said, would likely give people now in middle age—people such as himself—an additional thirty years of healthy life. Most of his fellow gerontologists thought this overly optimistic, though some had been persuaded of the value of his claims. "SENS 2.0" was where things crossed over into sci-fi territory—the longevity escape velocity theory, essentially.

"After those initial thirty years," he said, "the same people are going to come back looking for further rejuvenation. And the therapies, by that point, will have advanced significantly, because thirty years is a very long time in terms of any scientific endeavor. And so it is virtually one hundred percent certain that we will be able to rejuvenate those people even more effectively the second time than the first time. And so what that leads to is the idea that we'll be able to stay one step ahead of the problem indefinitely, to the point where we can treat people in such a way as they'll stay biologically in their twenties or thirties forever. Which translates very straightforwardly, at a conservative prediction, into four-digit life spans."

"Did you say four-digit?" I said, shunting my voice recorder across the table toward the extravagant edifice of his beard. "As in a thousand years?"

"Yes," he said. "Although that's, as I say, a conservative prediction. This is completely obvious, of course; it follows *absolutely* logically. The field of gerontology has started to come around to the idea that I'm right about regenerative treatments being the best way to postpone the effects of old age. But they don't want to risk their funding through any association with the notion of radical life extension, because it's perceived as total science fiction—even though, as I say, it's entirely logical. They find it absolutely necessary to distance themselves from this part of my vision."

"Just to clarify," I said. "I'm in my mid-thirties. What are my chances, would you say, of living to a thousand?"

"I would say perhaps a little better than fifty-fifty," he said, and drained the last of his beer. "It's very much dependent on the level of funding."

Aubrey excused himself to return to the bar, and as I sat alone at the table sipping my coffee, I attempted to assimilate the implications of what he had just told me. There was something familiarly unsettling about the movement of his logic, about the apparently rational means by which he reached what I could not help seeing as an entirely irrational conclusion. But my ignorance of the fields of genetics and gerontology precluded any adequate defense of my skepticism, and so it was not only out of mere politeness that I felt disinclined to inform Aubrey that what he was saying sounded, to my admittedly limited understanding, completely mad.

Aubrey returned, pint in hand. I told him, more or less flat out, that I was not convinced of the likelihood of him or anyone else coming up with a cure for death.

"Well, why not, eh?" he said. He narrowed his eyes over the rim of his pint, presenting me with the full magnitude of his gaze.

The problem with me, he said, was that I was far too willing to accept received authority, the opinions of so-called "experts," without examining the vested interests of those "experts," their need to say what they were saying–about his work, about the feasibility of radical life extension–even if it didn't necessarily accord with what they themselves believed. They dared not be seen holding controversial positions, he said, for fear of jeopardizing the flow of grant money toward their own work.

It was his belief that other gerontologists took note of what was said about him in the media, and made a conscious decision not to go near his work–to specifically avoid even reading it–because as scientists, they knew very well that they would, as he put it, "be unable to read something that is logically and cogently true without recognizing its truth."

The problem was not, in other words, that his fellow scientists found his claims ridiculous and false. It was far worse than that: they were afraid of being convinced of the truth of those claims and thereby coming to seem ridiculous themselves. And so what was preventing the mass of his fellow gerontologists from being persuaded by his work was, if I understood him correctly, precisely the irresistible force of its persuasiveness.

Such was the impenetrable circular system of Aubrey's self-belief.

Silicon Valley, with its "higher proportion of visionaries," was another matter entirely. The general cultural climate here in the Bay Area, the balmy atmosphere of technological possibility, was such that Aubrey's ideas had found a constituency, a place within a social context of radical optimism. (This latter formulation, incidentally, was one with which he took serious issue. "'Radical optimism'?" he said, reciting my phrase with theatrical derision. "*'Radical optimism'*? That sounds to me like you're saying overoptimism. And that is demonstrably not the case.")

When SENS relocated across the Atlantic, it set up its new headquarters just down the street from the Google campus in Mountain View—a proximity that presumably was more than mere chance. Life extension, a long-term preoccupation for Google's founders, Larry Page and Sergey Brin, had gradually become part of the company's "moonshot" culture. Google Ventures, the company's in-house corporate venture fund, was set up in 2009 under the leadership of a former tech entrepreneur named Bill Maris. Maris, who had said that he believed it possible to extend the life spans of people now living to five hundred years, and that he personally hoped to live long enough not to die at all, had invested heavily in biotechnology. (His friend Ray Kurzweil was hired by Google in 2012 in order, as *Bloomberg Markets* magazine put it, "to help Maris and other Googlers understand a world in which machines surpass human biology.")

When, in 2014, Google set up a new biotechnology firm called Calico—a research and development firm established with the goal of combating aging and age-related illness—Aubrey was exultant. Writing with characteristic grandiosity in *Time* magazine, he paraphrased

Winston Churchill: "Google's announcement about their new venture to extend human life, Calico, is not the end, nor even the beginning of the end, but it is, perhaps, the end of the beginning." He saw Page and Brin's decision to set up the company as a personal vindication, as well as an extremely encouraging sign that the war on aging was coming to be perceived as winnable. (Although, as he put it to me, if he were in Page and Brin's position, he'd "obviously have given the money to Aubrey de Grey.")

I left the bar. Out on Taylor Street, I glanced back through the window. Aubrey was still at the table, his laptop open now in front of him, his fingers moving at a rapid flutter across its keyboard. Against the noonday gloom of the bar, his face was lit by the soft glare of the screen, unreally white, and he had in that moment the strange luminescence of a medieval saint: the fanatical thinness, the holy fury in the eyes. I stood there looking at him for perhaps a minute, wondering what it might be like to believe in something as fiercely as Aubrey seemed to believe—to be so driven, so destined, so ordained. He didn't look up. He had, I supposed, already forgotten me.

In a 2011 *New Yorker* profile, Peter Thiel spoke about his investment in life extension research, his funding of projects like Aubrey's. Asked about the likelihood of such projects drastically exacerbating already dire economic inequality, given that the people most likely to benefit from them were the very rich, he said this: "Probably the most extreme form of inequality is between people who are alive and people who are dead." As with all advantages accrued by the wealthy, exemption from death would eventually trickle down, in some form or other, to the rest of us.

One of Thiel's more controversial philanthropic ventures was something called the Thiel Foundation Fellowship, through which he awarded gifted under-twenties $100,000 on the condition that they drop out of college for two years to focus on entrepreneurial activity.

In 2011, one of these fellowships was awarded to an especially brilliant MIT student named Laura Deming. Deming, originally from New Zealand, had moved to the U.S. at age twelve in order to work as a volunteer for the MIT biogerontologist Cynthia Kenyon, who became a long-term mentor. (Kenyon was then known for her 1986 discovery of a controlled mutation that increased the life span of the *C. elegans* nematode worm by a factor of six; by tweaking a single gene in the worm's DNA, Kenyon had enabled an organism with a natural life span of 20 days to live for 120 days, maintaining the level of vitality it ordinarily had at 5 days. In 2014, she became vice president of aging research at Calico.) At fourteen, Deming had enrolled as a biology undergraduate at MIT, and she was seventeen when she received the fellowship from Thiel, awarded to assist her in setting up the first venture capital fund directly focused on increasing human life spans.

The Longevity Fund, the VC firm which had resulted from that fellowship, was in its third year when I met Laura at her offices on the top floor of a lavishly nondescript building in Mission Bay. I was struck, initially, by the various ways in which she failed to conform to the stereotypes most people, myself included, would have in mind in imagining a venture capitalist focused on life extension. She was not, for instance, a middle-aged white American male who had amassed an immense fortune in technology and wanted to ensure an indefinite life span with which to enjoy the fruits of capitalism; she was, rather, a young woman of Asian descent who, despite enrolling in MIT at fourteen, did not conform to any geek stereotype I'd ever encountered. Laura's pleasantly businesslike and mildly self-deprecating manner was not quite successful in offsetting an imposing intellectual affect, which was all the more striking given the inescapable fact that the person sitting across a boardroom table from me was younger than many of the tersely hungover English literature undergraduates I'd taught over the years.

And so a strong cognitive dissonance arose from the three-way juxtaposition of Laura's extreme youth, her position in the business world, and the nature of her work; but it began to make sense within the context of the fact that she had been monomaniacally preoccupied with death for the past thirteen years.

"I have never not felt like extending the life spans of human beings is the correct thing to do," she said, measuring her words carefully. "When I was eight years old, my grandma came to visit us, and I remember wanting to play with her, and seeing that she wasn't capable of running around. And I remember realizing that there was something about her body that was, like, broken. And I thought, well, obviously somebody must be working on a cure for this disease that my grandma has. Then I realized that in fact nobody was working on any cure, because what was wrong with my grandma was not viewed as an illness. It wasn't even viewed as being *wrong*."

Not long after that, she came to understand that the way in which her grandmother's body was broken was merely an advance symptom of an absolute and final breakage, which would cause her to stop existing entirely. This unsettling insight into her grandmother's fate was quickly followed by the deeper recognition that this was, in fact, a universal phenomenon–*the* universal phenomenon–and that precisely the same fate therefore lay in store for her parents, and her friends, for everyone she knew and did not know, and for herself.

"I cried," she said, "for like three days straight."

Laura become obsessed with the idea of dedicating her life to addressing this unacceptable situation; by the age of eleven her ambitions were fixed on, as she put it, "starting a for-profit entity in the aging biology space."

She was wary of the term "life extension"; she used it a couple of times in our conversation, but then corrected herself, saying that she preferred to speak of "reversing the aging process," or "making people feel better while they're older." The problem with the term "life extension," she said, was that it evoked "insane people who have no scientific background convincing themselves that they will never die."

I got the sense from Laura that, for all the care she took in distinguishing her work from the more fantastic forms of techno-immortalism, she was shrewdly downplaying the extent of her own fixation on eradicating death.

A peculiar reality of modern medicine, she said, was the vast number of pharmaceutical companies pursuing treatments for cancer and diabetes and Alzheimer's—conditions that overwhelmingly resulted from aging—while virtually no companies were pursuing the underlying condition itself, which was the cellular degeneration of the human organism over time.

"I do believe," she said, "that death from aging is the biggest problem facing humanity. But I don't really talk about these things when I'm talking to people about investment, or about the VC fund. It has all the appeal of talking about a cult. People don't see the radical extension of life span as an investible model. It seems crazy if you haven't been steeped in the science, if you don't have a full understanding of the possibilities."

In terms of immediate investment prospects, Laura was especially excited by drugs that were already on the market. Diabetes treatments in particular, she said, tended to demonstrate an untapped potential for increasing the life span of organisms.

"There's this strange alignment," she said, "between insulin, blood sugar levels, and life span, and we haven't yet figured out why."

One drug Laura was especially excited by was a treatment for type 2 diabetes called metformin, which prevented the release of excess sugar into the bloodstream, and slowed the rate of cell turnover. It had been proven in tests, she said, to significantly expand the life spans of mice. Not long after we spoke, I read a news report on how the United States Food and Drug Administration had approved a five-to-seven-year clinical trial of metformin in humans, to be conducted at the Albert Einstein College of Medicine in New York, called Targeting Aging with Metformin (or TAME). I did a Google News search for the drug, and found an article in *The Telegraph,* which featured an interview with Laura—a "science wunderkind" who was "spearheading research into 'magic' anti-aging drugs." The headline, above a photograph of Laura performing tests in her laboratory, was a classic of the just-asking-a-question-here school of newspaper headline writing: "Could This Pill Be the Key to Eternal Youth?"

About a week after his third birthday, my son began to take an interest in the question of death. He began to take an interest, specifically, in the deaths of his mother and me. On hearing mention of my wife's grandmother, he had become immediately curious about who she was, and where she was. Not being religious, and not wanting to bullshit him, we felt we had little choice but to tell him that she was no longer around because she had died before he'd been born. He was already, at that point, familiar with the concept of death, but only really in the abstract and technical sense, as a thing that could happen, that *might* happen. We had introduced him to the concept, in fact, primarily as a means of discouraging him from running out in front of cars. If you got hit by a car, we told him, that would be it, the end of everything. He'd be gone.

All gone, we had said. Finished.

His cousins' dog, Woofy, had recently dropped dead from old age, but rather than trying to explain to him that she'd simply laid herself down on the kitchen floor and ceased to exist, we'd told him that she'd failed to be sufficiently careful, and had as a consequence been hit by a car. *Bang!* No more Woofy.

But now he wanted to know why his maternal great-grandmother was dead.

"Was she not careful?" he asked.

This made us laugh, a little, but it was, in the end, comprehensively unfunny. My wife's grandmother had been dead for years by then, and I had interacted with her only a handful of times while she was alive, yet I felt a faint and subtle swell of sadness now at her loss, the awful fixedness of it. I tried to remember what she looked like, but could do no better than conjuring a more or less generic image of an elderly woman. Low-sized, white-haired, glasses. A walking cane? The most extreme form of inequality.

We told him that it wasn't so much a matter of carelessness on her part as a result of her getting very, very old. When people got very, very old, we told him, they died.

And this was news to him. Dying, as far as he'd previously been aware, was what happened to people who got hit by cars, or–more excitingly–what happened to bad guys when good guys shot them.

He wanted to know whether we would get very, very old and die.

Because we felt we had no choice, we told him that we would, yes, eventually get very, very old and die–but not for an extremely long time. So long from now, we told him, that it wouldn't seem so awful when it did happen. But he was solidly against the idea from the first. He didn't want us to get very, very old and die. Not even eventually. Not even for an extremely long time.

One evening, my wife was putting him to bed, and he brought up the topic again.

"Do mamas and dadas really get old and die?" he said.

And my wife, feeling he had to be protected from this terrible awareness of how things were, told him that perhaps by the time he got to be Mama and Dada's age, there might no longer be any death, and so maybe he wouldn't have to worry about it at all. It was such a long way off, after all, and who knew what might happen between now and then. There were a lot of very clever men and women working very hard on the problem of death, she said, and perhaps they would succeed in solving it.

"You know how Dada has to go to America sometimes, for the book that he's writing?" she said.

"Yes," he said.

"Well, that's what Dada's book is all about. It's all about how in the future, when you're big, maybe nobody will ever have to die anymore."

We had no access to heaven, but this seemed a useful substitute. Not as powerful or evocative an idea, perhaps, but a release valve, nonetheless, for the psychic pressures of mortality. And it seemed to work. The problem of death had been solved, at least in our house, at least for the time being.

The Wanderlodge of Eternal Life

IN THE AUTUMN of 2015, a man of my acquaintance purchased a forty-three-foot recreational vehicle—a 1978 Blue Bird Wanderlodge, to be precise—and, having made to this vehicle such modifications as would lend it the appearance of a gigantic coffin, set out to drive it eastward across the great potbellied girth of the continental United States. His reasons for doing so were, in certain respects, complex and conflicting, but for now it will suffice to inform you that this voyage was undertaken in order to raise awareness of two distinct but related matters. The first of these was the regrettable fact of human mortality, and the need to do something about it; the second was that of his candidacy in the following year's presidential election.

This man's name was Zoltan Istvan, and I had known him for about a year and a half by the time he began his progress across the country—from the Bay Area, where he lived, to the Florida Keys, and thence northward to Washington, D.C., where he planned to ascend Capitol Hill and, in coy allusion to Martin Luther's delivery of his 95 Theses, affix a Transhumanist Bill of Rights to the great ornate bronze door of the Rotunda.

In an article on *The Huffington Post*, utilitarianly entitled "Why a Presidential Candidate Is Driving a Giant Coffin Called the Immortality Bus Across America," Zoltan had laid out his reasons for same. "I'm hoping," he wrote, "that my Immortality Bus will become an important symbol in the growing longevity movement around the world. It will be my way of challenging the public's apathetic stance on whether dying is good or not. By engaging people with a provocative, drivable giant coffin, debate is sure to occur across the United States and hopefully around the world. I'm a firm believer that the next great civil rights debate will be on transhumanism: Should we use science and technology to overcome death and become a far stronger species?"

I had first met him at the conference in Piedmont. It was Hank Pellissier who had introduced us. He was constructed on a noble scale, Zoltan, handsome in an irrefutable and yet somehow unserious fashion—like a life-size Ken doll, or a proof-of-concept for an Aryan eugenic ideal. I recognized right away that he was not a typical transhumanist. He was polite and charismatic, and in no obvious sense geeky or awkward.

He gave me a copy of a book he'd recently self-published called *The Transhumanist Wager,* an unwieldy novel of ideas about a freelance philosopher named Jethro Knights (a character with certain key biographical particulars in common with his creator) who sails around the world to promote the need for life extension research, and winds up establishing a floating libertarian city-state called Transhumania—a haven for unhampered scientific research into human longevity, a regulation-free utopia of tech billionaires and rationalists—from which he wages an atheist holy war on a theocratic United States.

A couple of days later, at a café in San Francisco's Mission District, he told me of how the novel had not gone over well with any of the 656 agents and publishers he'd sent it out to over the previous year. He'd spent over a thousand dollars, he said, on postage alone. Self-publishing had been the only available option, but he was pleased with how the book was selling, the impact it was having within the transhumanist movement. It was, he said, starting conversations. The book's cover, which he himself had designed, featured a greenish negative photo of his own face, in profile, staring into the vacant sockets of a human skull. He himself was the first to acknowledge that it was not entirely successful, from an aesthetic point of view.

"It's supposed to be like Hamlet," he said. "You know, with the whole Yorick scene? With me facing the prospect of death and all that? But yeah, I'm not sure it really works."

I glanced down at the book in front of us on the table, and did not disagree. We were sitting in a courtyard at the rear of the café, in blinding midday sunlight. The tables were all occupied, I noted, and yet we were the only people conducting a conversation. Every other customer in the café was alone, and typing on an Apple laptop. As so often in San Francisco, I had a sense of being embedded in some hyperreal simulation of a corporate utopia—or, rather, a heavy-handed parody of such a thing. As a scene, it felt a little overbearing in its symbolism. This is one of the problems with reality: the extent to which it resembles bad fiction.

"I've seen worse book covers," I said, which for all I knew might have been the truth.

With Zoltan, the impression was of a man who, having reached his early forties, was trying to recapture the existential vitality of his youth. In his twenties, having graduated from Columbia University with a degree in philosophy, he'd fixed up an ancient yacht and, alone but for the dozens of nineteenth-century Russian novels with which he'd stocked the vessel, set out to circumnavigate the globe. He had partially funded his trip by making short documentaries for the National Geographic Channel about the remote places he was visiting. Somewhere along the way, he had invented an extreme sport called volcano boarding (basically the same thing as snowboarding, except you did it on the slopes of an active volcano). While reporting on the large number of buried land mines still remaining in Vietnam's DMZ, Zoltan himself came very close to stepping on one of these devices; his guide tackled him from behind as he was walking, and brought him to the ground just inches from where an unexploded mine was jutting from the earth.

In the narrative he had constructed about his life—his origin story—this was the moment he'd become a transhumanist, the moment he'd become consumed by an obsession with mortality, with the unacceptable fragility of human existence. He returned to California to set up a real estate business and, taking full advantage of the permissive finance culture of those years, bought and flipped a number of properties in quick succession. He hated the work, but he was good at it, and made quite a lot of money very quickly. Right before the crash in 2008, he sold half his portfolio, and came out of the deal a millionaire. He hung on to the rest of it, which included a number of houses on the West Coast, some land in the Caribbean, and a straight-up vineyard in Argentina. Forty years after his own parents had fled the Communist People's Republic of Hungary, he was the embodiment of an American capitalist ideal: the immigrant son with a weird European name who'd become an honest-to-God self-made millionaire. It hadn't even been that hard. The system worked. Money worked.

And that money was enough to enable him to quit his job, and to dedicate himself to several years of writing *The Transhumanist Wager,* into which project he channeled all his ideas about the possibility, and the necessity, of achieving physical immortality through science.

That day in the Mission, Zoltan told me of how his wife, Lisa, a gynecologist who worked for Planned Parenthood, had recently started to express a keen interest in his doing something productive with his life. Lisa had, at that point, just given birth to their second child, and, what with the exponentially growing cost of living in the Bay Area, and Zoltan's reluctance to sell any further properties, she was becoming increasingly concerned about the need to begin saving for their two daughters' educations. He himself, he said, was reluctant to fritter away money on such things, given that by the time the girls were in their late teens it would be possible to upload the informational content of a Harvard or Yale degree directly to their brains anyway—and at a fraction of what such an education costs today.

Lisa, he said, was largely tolerant of his views, but drew the line at gambling their children's futures on the fanciful notion of some imminent technological intervention.

"Obviously she's a little resistant to transhumanist ideas," he explained, "because in the near future her entire profession will be obsolete. What with actual childbirth becoming a thing of the past. You know, with babies being produced by ectogenesis and whatnot."

"Your wife sounds like a smart lady," I said.

"Oh, she is," he said, finishing his latte. "A very smart lady."

When, some months later, Zoltan emailed me about his decision to run for president, I immediately called him. The first thing I asked was what his wife thought of the plan.

"Well, in a way," he said, "it was Lisa who gave me the idea. Remember how I said she wanted me to do something concrete, get some kind of a proper job?"

"I do," I said. "Although I'm guessing running for president on the immortality platform was not what she had in mind."

"That's correct," he confirmed. "It took a little while for her to come around to the idea."

"How did you break it to her?"

"I left a note on the refrigerator," he said, "and went out for a couple hours."

I will admit that I was never as appalled by Zoltan as I should have been. I will admit, that is, that I liked him more than I had any business liking him. This strikes me as an important point, and maybe even a structural one. He was, in many ways, an embodiment of everything that was most questionable about transhumanism—of its extremity, its blindness to human nuance, to anything but the most bluntly instrumental measures of human value.

Once, he told me about a thing that had happened in a coffee shop he frequented in Mill Valley, the upscale North Bay neighborhood where he lived with his family. He'd gone to the coffee shop to get out of the house for a bit, and to do some work on his laptop. A man and his teenaged son came in, and the son, who was profoundly mentally handicapped, slipped free of his father's grasp, and began to run around the coffee shop, bumping against tables and knocking things over. One of the tables the boy bumped against was Zoltan's, causing his coffee to spill over his laptop.

The point of the story, as always with Zoltan, was to suggest that technology might be brought to bear on such regrettable human situations. The incident had got him thinking about whether it might not be sensible–better, on the whole, for the boy, and for his parents, and for society at large–that such profoundly afflicted people be cryonically suspended early in life, put on ice, as it were, until such time as we possessed the technology to cure them of their conditions.

The laptop, incidentally, was unharmed.

"The question to ask," he said, "is this: If you were that individual, would you want that done to you? Would you want to live a life where you can't think, where you're just running around crazy all the time? Or would you want society to just rise up and say, you know what? Of course this is very challenging ethically, but we believe that in fifty years we will have the science to make this person all that he can be. So if we cryonically suspend him now, we will be giving him the prospect of a normal life in the future."

This view seemed a consequence of transhumanism's extreme instrumentalism, a view of life in which intelligence, pure use-value, took precedence over all other concerns. (I thought of Tim and Marlo, of Anders Sandberg and Randal Koene: their ecstatic visions of ascension to pure mind.) Zoltan saw the boy in his story as a broken machine, a mechanism serving no purpose to himself or others, but whom technology might be called upon to put to rights, to *save*. It is important to understand that the intended implication of Zoltan's story was one of optimism. Zoltan was nothing if not an optimistic man.

It struck me that there was something grandly and irreducibly American in the specific gesture of running for president, in the notion that–if only in theory, if only in symbol–it was the right and the possibility of every individual to seek absolute power, absolute influence, in the name of some cause, in the name of himself.

It should be said that although he was a highly ambitious man, Zoltan's decision to run for president was by no means motivated by a delusion that he would ever make a significant impact at the ballot box, if he ever managed to get that far. It was motivated by another manifestation of unrestrained optimism: that death was a problem to be solved. That we could, all of us, be put to rights by technology.

And this went to the core of what I found so compellingly strange about transhumanists as a whole, about their values and motivations: the idea that, as a culture, as a species, we needed to be shaken out of our complacency about mortality. Not in the existential sense that we should live with an abiding consciousness of death's inevitability, but in the precisely opposite sense that such a belief in death's inevitability was itself a form of complacency, an excuse not to address the problem.

I wanted to get as close as possible to this idea, to follow it as it made its way from the Bay Area to the American heartland. And so I made plans to get on the bus.

By the time I joined him on the campaign trail, in late October of 2015, Zoltan's fortunes had in many respects changed for the better. Due to the media interest his presidential bid was accruing, he was now one of the most prominent figures in the transhumanist movement. Documentary crews from Vice and Showtime had recently been following his progress across California and Nevada, and his personal brand had seen a considerable uptick in value: he had lately entered the promised land of the lucrative corporate speech, the $10,000 conference appearance fee.

That a man who had until so recently been an unknown quantity had risen so swiftly to prominence, and that, because of his self-proclaimed leadership of the Transhumanist Party, the media had begun to identify him as a de facto leader of transhumanism per se: all of this was causing considerable disquiet within the movement. Among the old guard, there was a growing sense of Zoltan as a usurper, as a man who had come from nowhere and had hijacked the movement, and the term "transhumanism," for his own ends.

When I joined him on a Friday morning in Las Cruces, New Mexico (the plan being to drive the bus across Texas to give a campaign speech at a biohacking event in Austin the following Monday evening), he had just driven from Phoenix, where he'd met with Max More at Alcor. It was a meeting he'd anticipated with some anxiety, I knew, because Max, along with several other members of the transhumanist old guard—"the elders," as Zoltan referred to them with no apparent irony—had signed a petition disavowing his presidential campaign, disassociating themselves from him and his party, which, as Zoltan himself was forced to acknowledge, was not really a party in any meaningful sense anyway, so much as just him and a handful of advisors. (At that point, Aubrey de Grey was on board as the campaign's "anti-aging advisor," and Martine Rothblatt's son Gabriel—who had himself run for Congress in 2014—was acting as political advisor.)

That morning, at my hotel in El Paso, I'd gone online and watched a report on Phoenix's Channel 3 News about Zoltan's campaign stop in the city. "The *plan*," drawled the reporter, "is to drive this *sarcophagus* on *wheels* to Washington, D.C., and to convince the White House, as well as *Congress*, to put more money into *immortality* research." The report had covered his visit to Alcor and, from what I could see, it seemed cordial enough.

I met Zoltan outside an empty secondhand bookstore on Las Cruces's main street. His hair was neater, blonder, than when I'd last seen him seven months previously, and his face and neck were mottled from exposure to the desert sun. He was accompanied by an exceptionally tall and willowy young man with long black center-parted hair and wide, ascetic eyes. This young man held a video camera attached to a tripod in one hand; the other he extended toward me in solemn salutation.

"Roen Horn," he said. "Do you want to live forever?"

"I'm not sure that I do," I said, feeling the slender bones of his hand as I received it in my own.

"Well, why not?" he said. "Do you want to die? Do you think death is a good thing?"

"These are tricky questions," I said. "Can I think about them on the bus and get back to you?"

This Roen Horn, I learned as we walked down the eerily deserted main street, was a volunteer for Zoltan's campaign, a zealous advocate of radical life extension who was also making a documentary about the Immortality Bus. This latter vehicle of transcendence was currently moored in the parking lot of a nearby Bank of America. The immediate plan, Zoltan told me, was to drive out into the desert to White Sands Missile Range, the largest military installation in America, where he intended to stage a protest highlighting the need to divert public money away from weapons spending and into life extension.

The Wanderlodge was an even odder spectacle than I had anticipated–a great brown absurdity with the words "IMMORTALITY BUS WITH TRANSHUMANIST ZOLTAN ISTVAN" neatly hand-painted in white across the length of its midsection. On the rear of the bus were painted the words "SCIENCE VS THE COFFIN." To the roof was affixed a construction of inward-slanting wooden boards, likewise brown, on top of which in turn rested an elaborate arrangement of synthetic flowers. The effect of all this was not un-coffin-like, but it helped to know what you were supposed to be looking at.

Within were all the trappings and creature comforts of a mid-range 1970s bachelor pad: a kitchenette equipped with ice machine and microwave oven, a dining table, ample bench-style seating for on-road lounging, and, toward the rear, two narrow bunks and a bathroom (nonfunctional). Orange shag pile carpeting featured throughout.

The thing was roadworthy, more or less—as long as you didn't drive it uphill at too steep a gradient, and as long as you pulled in every ninety minutes or so to change the engine oil, which leaked out the side at a truly dramatic rate. This steady leakage was a concern for Zoltan, not just with respect to the long-term health prospects of the Immortality Bus, but more urgently the likelihood of our getting pulled over on the freeway by a traffic cop, something that, given the conspicuousness of the vehicle, seemed a nontrivial prospect.

The difficulties began about half an hour outside Las Cruces. As the freeway slung a wide loop around the jagged foothills of the Organ Mountains, the sound of the engine, striving to haul us uphill, had become an alarmingly shrill rasp. We were maxing out now at about 35 mph, and Zoltan's hulking form was bent low over the wheel, as he eyed the dashboard's archaic array of mysterious dials.

"We look to be overheating pretty bad," he said. "I've never seen it this far into the red. And this isn't even a particularly big hill. We could have a problem here, gentlemen."

It was Zoltan's custom to address Roen and me, collectively, as "gentlemen"—a verbal gesture that was more comradely than formal.

Upward trajectories were best avoided, he explained, on account of a cruel little paradox at work in the ancient mechanics of the Wanderlodge: the longer you went uphill, the harder the engine had to strain to move the bus at even a sluggish pace; and the slower you went, the less air circulated from outside to cool the engine, thereby perpetuating the vicious circle of overheating.

Another, simpler, way of putting this would be to say that the radiator fan was fucked.

We crested the hill and began to pick up speed on the downward slope. The engine's whining descended somewhat in pitch, and I was newly confident that we were no longer about to grind to a halt in the desolate heat of the desert.

"That's a relief," I said.

"Actually," said Zoltan cheerfully, "it's far more dangerous coming downhill, because we're relying on forty-year-old brake pads here.

This bus, you have to just take it slow, because there's no way to guard yourself against bad brakes."

In the light of this new information, I felt we could have been taking it a fair amount slower than we were. I recalled with some discomfort that the man driving the vehicle had invented the sport of volcano boarding, presumably as a way of solving, in one deft move, the problems of the insufficient riskiness of both snowboarding and hanging out on the slopes of active volcanoes. Although I was not sure that I wanted to live forever, I was sure that I didn't want to go down in a blaze of chintzy irony, plunging into a ravine strapped into the passenger seat of a thing called the Immortality Bus.*

Between the driver and passenger seats there was a large, raised, shag-pile-carpeted area, which I was using to lay out my various writ-

* Some version of this thought occurred to me several times a day throughout my time on the bus. For all that Zoltan railed against the tyranny of death over human lives, and for all that his political platform rested on the conceit that physical immortality was within our grasp as a species, his attitude toward basic road safety was at times wildly cavalier. The fact that he was piloting a forty-three-foot coffin bus through West Texas did not, for instance, stop him from checking his phone every couple of minutes–responding to texts and emails, checking the social media analytics on his latest piece for *TechCrunch*, etc. There was even, one evening, a brief and spirited burst of drunk driving, from the parking lot of a Fort Stockton fast-food restaurant (where he'd enjoyed several generous whiskeys with his evening meal) to our nearby motel. This infraction–of both federal law and the spirit of life extension–he justified by pointing out that the steering on the bus was so unresponsive that even if your driving was somewhat erratic, the vehicle's trajectory tended to remain relatively steady anyway. For his part, for a guy who was motivated in everything he did by an overwhelming terror of death, Roen seemed weirdly reluctant to put a seat belt on. He spent most of our road trip lying faceup on the couch behind the driver's seat, which position seemed to me to fly in the face of not just basic road safety, but also his own frequently stated life goals. I found myself thinking of a *New Yorker* profile of Peter Thiel that noted his failure to wear a seat belt while driving his sports car around the freeways of the Bay Area. All of this seems especially odd, given the seat belt's well-known efficacy as a life extension technology.

erly impedimenta–voice recorder, notebook, pens, and so forth. This, it turned out, housed the Wanderlodge's actual engine. At one point, Zoltan decided that it might ameliorate the overheating issues if he opened this up to "let the engine breathe." Due to its busted air-conditioning unit, the cabin of the Wanderlodge was already pretty hot, but as soon as we lifted the lid of this housing, the whole interior was quickly transformed into a sort of hellish, hurtling sauna, heated by the searing petroleum fumes that emanated from the roaring shag-pile maw of the open engine unit.

I unbuckled myself from the passenger seat, and went to sit on one of the couches, where the blast of heat and smoke from the engine compartment was slightly less intense.

"I KNOW IT'S NOT PLEASANT!" bellowed Zoltan affably over the near-deafening roar of the engine, "BUT IT'S REALLY HELPING WITH THE OVERHEATING!"

At length we pulled in to let the engine cool awhile, and Zoltan went outside to change the oil. Roen was recumbent on the long couch to the rear of the driver's seat, staring impassively at the ceiling, his hands cradling the back of his head. This was to become his default attitude throughout the trip.

I craned around in my seat and asked him how he'd wound up volunteering for Zoltan's campaign.

"I just really don't want to die," he said. "I can't think of anything that would suck more than being dead. So I'm just doing what I can to ensure that life extension science gets the funding it needs."

"So what is it you do?"

"What do you mean?"

"I mean work-wise. When you're not volunteering for Zoltan."

"I run the Eternal Life Fan Club," he said. "It's an online organization for people who are serious about living forever. Not, like, five hundred years like a lot of transhumanists. *Forever.*"

Like many transhumanists, he was deeply convinced of the importance of Aubrey de Grey's SENS project. Aubrey was, for Roen, a figure of near-messianic dimensions. Most of what little money Roen raised as a life extension advocate went to supporting SENS.

He was, too, a huge fan of Laura Deming. When I said that I'd met her, he reacted as though I'd mentioned a film star.

"She's a hero to me," he said. "I love her. She's out there fighting against death. I use a lot of her quotes as memes."

He opened up his laptop and clicked around a while and, by way of evidence, presented me with an image of Laura, posted to his Facebook page, with a quote beneath it: "I want to cure aging. I want to make us all live forever."

Roen was twenty-eight, and lived in Sacramento with his father, a recently retired insurance claims adjuster, and his mother, who worked in a movie theater. His parents were devout Calvinists who believed in eternal life in paradise for the elect, and in eternal damnation for the unchosen. His father, who was especially hard-line, was vocal in his conviction that his atheist son was destined for the infernal torments of hell.

"How does he feel about this whole Immortality Bus thing?" I asked.

"He's actually okay with it," said Roen. "He thinks it's cool I'm getting to be on TV news and all."

The military testing area at White Sands, New Mexico, is a forlorn and silent place, unfurling eastward from the Organ Mountains into the desert solitudes of the Tularosa Basin. It was here that the boundaries of technological possibility, the boundaries of fear, were redrawn by men of science in the last days of the Second World War; it was here, in July of 1945, that the first atomic bomb was detonated, the prototype of the Fat Man plutonium device that was two weeks later delivered from the heavens to the mortals of Nagasaki.

Just past the security checkpoint at the entrance to the facility, there was a kind of open-air munitions exhibit that featured a squat replica of the Fat Man, along with dozens of other decommissioned rockets and bombs. In the undulating heat of the desert, these slender tilted obelisks loomed like the inscrutable monuments of an ancient thanatopia, a henge of metal phalluses thrusting skyward in ecstatic communion with the cosmic powers.

Zoltan removed from his backpack a banner he'd had printed for the occasion and, positioning himself in front of one of the larger rockets, instructed Roen to take a series of photographs of him bearing the unfurled message: "TRANSHUMANIST PARTY PREVENTS EXISTENTIAL RISK." The intention of the protest, such as it was, was to create a series of images and short videos to be uploaded to Zoltan's various social media accounts and shared among his many thousands of followers. It was a self-conscious simulacrum of protest; it was politics as content, content as pure form.

Leaning self-consciously against the Fat Man replica, I scribbled in my notepad. Roen took out his phone, and filmed a six-second Vine video of Zoltan saying "Stop nuclear war! It's a devastating existential risk!" Then he filmed Zoltan giving another brief speech on the central theme of his campaign: the need to divert government spending away from war and into research on life extension.

In my notebook, I scrawled Oppenheimer's famous quotation of Vishnu, preserver of the cosmos: "Now I am become death, destroyer of worlds."

Here at White Sands, science made its nearest approach to divine likeness, divine knowledge. Here, with these experiments in celestial violence, humanity came closest to transcending itself, fulfilling itself.

It was Oppenheimer who gave these nuclear tests the code name Trinity. Asked, years later, why he had chosen this theological designation, he said that he was not entirely sure, but that he felt it had something to do with his love of the metaphysical poetry of John Donne.

Later that evening, we pulled in off the Interstate and checked into a motel; I stood in the doorway waiting for Zoltan and Roen to get their stuff from the Wanderlodge, and browsed through a stand of leaflets by the entrance. Most of these advertised sites of general touristic interest—the International UFO Museum & Research Center at Roswell, for instance, and PistachioLand, "home of the world's largest pistachio."

There was also a small assortment of Christian pamphlets, and of these I selected one that was simply entitled "Eternity." It was a prospectus of the apocalypse, published by an outfit called the Gospel Tract and Bible Society. Standing in the empty lobby of the motel, I read of God's decree that all things shall cease to exist—that "the heavens shall pass away with a great noise, and the elements shall melt with fervent heat, the earth also and the works that are therein shall be burned up"—and I thought again of the unearthly monument I had walked around that day, the ceremonial circle with its ranged machineries of death.

I read on, and learned of how I, or my soul, might survive the death of my body and all other worldly things by submitting myself entirely to the Lord. "In all of creation," I read, "only man, clothed with a changed and immortal body, will make the transition from time into eternity. Man is the only creature who has the 'breath of life' (Genesis 2:7), which lives forever as God does."

I remembered asking Roen, earlier that day, about how his evangelical upbringing might have informed his belief that he would live forever through science. He'd said there was no longer any need for gods.

"Science is the new God," he had said. "Science is the new hope."

The Immortality Bus continued its slow, straining progress toward Austin. Now and then we passed a hand-painted sign standing in a field, a gesture of anonymous pride or defiance. A MAKE AMERICA GREAT AGAIN: DEPORT OBAMA. A DON'T MESS WITH TEXAS. Mostly, we passed roadkill. For miles at a stretch, the only landmarks were corpses—foxes, raccoons, armadillos in various states of putrefaction on the margins of the Interstate.

"Dead animals everywhere," I scrawled in my notebook. *"Vultures omnipresent. (Over-literal?)"*

Both Zoltan and Roen had grown up in deeply devout families, Catholic and Calvinist respectively. Their fervent atheism, their rationalist zealotry, was both an effacement and a continuance of those religious backgrounds. Theirs were souls on fire for science, ablaze with the love of reason and all its works.

But the cold insistence of science was that nothing was permanent, nothing would last, that everything was ultimately roadkill, including the road itself. The second law of thermodynamics insisted that the universe was in a state of ongoing and nonnegotiable decline. The pen I held in my hand, I noted, was running out of ink. The body with which I moved it was being drawn slowly but inexorably toward death. The Immortality Bus was literally falling apart. The cold insistence of science was that America would not be made great again, and that the sun would one day explode and engulf the Earth, and everything would be vaporized, and Texas would finally and irrevocably be messed with.

The earth also and the works that are therein shall be burned up.

This belief that science would offer us an exemption from our place in this vast panorama of disintegration—of which the rotting armadillos and raccoons, the circling vultures, were only the most immediate manifestations—was a displacement of a fundamentally religious instinct. I thought of the psychoanalytic concept of transference, whereby the patient's childhood relationship with his or her parents was redirected onto the figure of the analyst. Wasn't transhumanism precisely that: a wholesale projection of the formative relationship with God onto the figure of Science? Wasn't all of it—brain uploading, radical life extension, cryonics, the Singularity—a postscript to the oldest of narratives?

I wrote in my notebook: *"All stories begin in our endings."*

—

Roen's austere calorie-restrictive diet, oriented as it was toward maximum longevity, was ill-served by the truck stops and gas stations and drive-thru hamburger repositories of West Texas. And his abstention from alcohol and all other drugs seemed at odds with the general vibes he emanated, his wide-eyed and dreamy affect having given me the initial impression of an archetypal So-Cal stoner type.

I was coming to see him now as a transhumanist ascetic, a young man who had largely withdrawn from the world so that he might never have to leave it.

He was a figure out of Dostoevsky. He was, specifically, Alyosha Karamazov, of whom we are told the following in the early pages of *The Brothers Karamazov*: "As soon as he reflected seriously he was convinced of the existence of God and immortality, and at once he instinctively said to himself: 'I want to live for immortality, and I will accept no compromise.' "

I learned that, in his parents' house in Sacramento, Roen slept on the floor of his bedroom, partly because he didn't want to buy a bed when what little money he had might better be spent on supporting life extension research, but mainly because of an obscure hostility to soft surfaces. (This self-avowed aversion was roundly contradicted by his near-fanatical dedication to couch-based recumbence, as outlined above.)

We pulled in at a truck stop some hours west of Fort Stockton, took a booth at an all-you-can-eat buffet. At the table next to ours, an immense man, vastly Stetsoned, sat hunched over a Bible, opened to the Book of Job, while working his way methodically downward through a pile of miscellaneous viands, a flourishing ecosystem of meats and slaws and assorted carbs. While Zoltan fielded a call from his irate wife about an overflowing toilet he had failed to repair before setting off across the country to promote immortality, I took the opportunity to quiz Roen about his lifestyle choices.

"I have to admit," I said, "I find this whole immortality thing difficult to get behind. Doesn't your obsession with living eternally actually amount to your being totally imprisoned by death?"

"Maybe," he said. "But aren't we all? Isn't that kind of the whole idea?"

I told him that I took his point, and we both laughed, a little awkwardly perhaps, and ate our lunches in silence for a while, listening in on Zoltan's terse exchange with his wife.

Roen ate with ruminative care, as though apportioning to each mouthful of salad the optimal measure of mastication. Aside from being a strict vegetarian, he seemed to keep his dealings with food to a bare minimum. He rejected meat solely for health reasons, but I couldn't help but wonder whether on some deeper level it was also a further manifestation of his rejection of death itself, of the animal nature of his body.

"What are we to make," asked the psychoanalyst Ernest Becker in his book *The Denial of Death*, "of a creation in which the routine activity is for organisms to be tearing others apart with teeth of all types—biting, grinding flesh, plant stalks, bones between molars, pushing the pulp greedily down the gullet with delight, incorporating its essence into one's own organization, and then excreting with foul stench and gasses the residue. Everyone reaching out to incorporate others who are edible to him."

It was a deathly business, being alive, being an animal. Nature, for want of a better word, was evil.

It was late October, and the truck stop was lavishly bedecked with the ghoulish paraphernalia of the season—with miniature plastic jack-o'-lanterns, cotton spiderwebs, wall-mounted witches on broomsticks, and other festive gewgaws. Dangling from the ceiling directly behind Roen's head was a rubber statuette of Death himself, his skeletal form shrouded in a ragged black cowl, a plastic scythe clutched in his bony little hand. This cartoonish figurine twirled slowly on its nylon string, distracting me with its overblown enactment of bargain-basement ironic foreshadowing.

"I just want to have fun forever," said Roen at length, guiding a forkful of dry salad leaves toward his pale face. "The twenty years I get from eating the way I do could be the difference between my dying and my getting to longevity escape velocity. I'm holding off on pleasure now so that I can have more pleasure later. I'm actually a total hedonist."

"You don't seem even slightly like a hedonist to me," I said. "You don't drink, you don't take drugs. You barely eat. To be honest, you seem like a medieval monk."

Roen cocked his head to one side, gave the idea some consideration. I didn't want to raise the topic of sex, but it seemed to hang there, twirling slowly above our heads like the rubber avatar of Death. I didn't have to, as it turned out: Roen himself brought it up, after his own fashion.

"You know one really cool thing about being alive in the future?" he asked.

"What's that?"

"Sexbots."

"Sexbots."

"You know, like AI robots that are built for having sex with."

"Oh sure," I said. "I've heard of sexbots. It's a nice enough idea. You really think that's going to happen, though?"

"For sure," said Roen, closing his eyes and nodding beatifically, in momentary reflection upon some distant exaltation. "It's something I'm very much looking forward to."

He had a particular way of smiling that was half evasion and half challenge. Out of context, you'd maybe be tempted to describe it as smug, but the effect was somehow deeply endearing.

"The problem I have with sexbots," I said, "is why wouldn't you just have sex with an actual person? I mean, all things being equal."

He said, "Are you kidding me? A real girl could cheat on you, sleep around. You could get an STD. You could maybe even die."

"Is that potentially a bit alarmist?"

"No way, man. It happens literally all the time. See, a personal sexbot would never cheat on you, and it would be just like a real girl."

He said nothing for a time, and drank at leisure from his glass of water. He consumed some further forkfuls of salad. He gazed out the window at the parking lot full of trucks, the Interstate beyond, the ever-present vultures hanging in the air.

I said, "Do you mind me asking if you've had bad experiences with people cheating on you?"

He said, "I have so far abstained from sex. I have never had a girlfriend."

"You're saving yourself for the sexbots?"

He nodded slowly, shrewdly raising his eyebrows. You bet your ass he was saving himself for the sexbots.

"Fair enough," I said, raising my hands in affable capitulation. "I hope you live that long."

He said, "I'm pretty sure I will."

The widening rift between Zoltan and the "elders" was becoming an increasingly predominant topic of conversation on the bus. It was all quite a complicated situation, and there seemed to be several distinct factors in play. In an interview about his campaign for the website Vox, Zoltan had signaled his intention to abandon his campaign at some point before the election and to endorse the candidacy of whoever wound up being the Democratic nominee. Hank Pellissier, one of Zoltan's earliest supporters, had resigned as secretary of the Transhumanist Party in protest against this revelation, which he claimed was the "last straw."

Hank's defection had ignited further dissent from transhumanists who had all along been quietly unconvinced by Zoltan's campaign. Among these was Christopher Benek, a Florida Presbyterian pastor who was one of the most prominent Christian transhumanists, and who had until recently been on ecumenical terms with Zoltan. (Reverend Benek had raised some futurist eyebrows in 2014 by publicly suggesting that advanced forms of artificial intelligence should be converted to Christianity, reasoning that any autonomous form of intelligence was one that should be encouraged "to participate in Christ's redemptive purposes in the world.") Writing in *The Christian Post*, he had taken issue with Zoltan's "ideological tyranny," and his "imperiously appointing himself as all U.S. Transhumanists' representative," furthermore characterizing his run as "merely an attempt to globally claim transhumanism as a primarily Atheistic venture which openly rejects organized religion and God."

Further disquiet had been caused by Zoltan's announcement on Facebook of his intention, once he'd finished his presidential campaign, to establish "a global political party that embraces the idea of being the leading voice and influence in a worldwide government." Zoltan had always been vocal about his belief in the abolition of national borders, but he seemed now to be following his libertarian logic to some paradoxically authoritarian ends. It was hard to imagine this surprising anyone who had read *The Transhumanist Wager*, but the announcement served to further alienate all but the most extreme techno-rationalist of Zoltan's supporters.

And then had come the petition to disavow his campaign, whose signatories rejected both Zoltan and his Transhumanist Party, "so long as it cowers under authoritarian control, so long as it denies the diversity of Transhumanist values, and so long as it mongers unnecessary hostility toward others."

His increasing tendency to publicly adopt outlandish political positions was a major factor of this growing dissent. That spring, for instance, he had published an opinion piece on Motherboard, Vice's technology website, arguing that the $1.3 billion recently budgeted by the city of Los Angeles to make its streets and accessways more wheelchair accessible would be far more sensibly spent on investing in robotic exoskeleton technology. "Let the sidewalks remain in disrepair," he wrote. "Instead, in the transhumanist age we're now in, let's work to repair physically disabled human beings, and make them mobile and able-bodied again."

When I discussed this with him, Zoltan seemed genuinely not to understand why people with disabilities were so deeply offended by this insistence that it was they who needed "fixing," rather than the discriminatory attitudes that were manifested in the urban environment, and in comments like his own. The underlying premise of transhumanism, after all, was that we *all* needed fixing, that we were all, by virtue of having human bodies in the first place, disabled. (I found myself recalling, here, Tim Cannon's assimilation of the language of transgender experience to the transhumanist context—his insistence that he was trapped in the wrong body by virtue of having a body at all.)

Unchastened by the exoskeleton debacle, Zoltan later went on to suggest that an elegant solution to the debate over the Obama administration's plan to take in ten thousand refugees from the Syrian civil war would be to implant those refugees with microchips as part of their admittance process. Such a policy, he claimed, would enable the government to track their movements, to determine whether they were plotting terrorist atrocities, and "monitor whether they were contributing to the system, paying taxes or causing strife." He was aware of the extent to which people found this idea repugnant, but again seemed basically untroubled. His response to concerns about this advocation of unprecedented intrusion of government into the lives—into the very bodies—of human beings was to say that "maybe Big Brother isn't the bad guy, if he protects us from ISIS." Besides, he himself had had an RFID chip implanted at a grinder event earlier in his campaign tour, and the procedure had been much less painful than you'd think. Once the refugees were deemed not to be a threat to public safety—after a probationary period of, say, three years—they might even choose not to have their microchips removed, given that the technology would soon enable them to pay for coffee in Starbucks by waving their hands at a chip reader.

To the extent that all this was motivated by any ideology to speak of, it seemed to me to be motivated by an ideology of technology itself: an imperative to increase the merger of humans and machinery by whatever means necessary. (As such, Zoltan often seemed to me like a walking illustration of Theodor Adorno and Max Horkheimer's argument, in *Dialectic of Enlightenment,* that the progress of scientific rationalism was always a progress toward tyranny. "Technical rationality today," as they put it, "is the rationality of domination. It is the compulsive character of a society alienated from itself.")

In his more expansive moments, he spoke of the possibility—"if this trajectory we're on holds"—of his eventually surpassing Kurzweil in terms of influence. "I can bring a huge amount of younger people into transhumanism," he said. "I'm actively trying to build a movement with these millennials, so they'll change the culture." He was obsessed with impact, with eyeballs; he spoke of retweets and engagements and Facebook likes and other metrics as though such things were the true currency of the new world, and he made it clear, time and again, that the "elders" could not hope to match his impact in this sphere. The media loved him, and he loved that they loved him, and he loved that the former leaders of the transhumanist movement hated that they loved him.

I was impressed by the multidisciplinary sweep of his ambitions, his almost mystical conviction about his own rise to greater heights of influence and power. Frequently, he spoke of the environmentalist movement as a model for how he planned to build transhumanism, and radical life extension particularly, into something the general public, and eventually governments, would be forced to take seriously. It was clear that he thought of himself as the Al Gore figure in this model.

My feelings toward Zoltan were complicated, contradictory, and subject to sudden mutations, intensifications, reversals. His grandiosity exerted a paradoxical magnetism, tempered as it was by an easygoing self-deprecation. He would be talking about wanting to change the world, convincing folks that physical immortality was within their grasp, and, in the next moment, he would be taking ironic delight in some scheme he'd come up with to keep the Wanderlodge on the road for another few hours.

"That's what I'm good at, half-assing stuff," he told me one afternoon in the parking lot of a Walmart where we'd stopped to fill a cart with containers of engine oil, and some barbecue trays to collect it as it leaked from the bus.

I said I'd begun to think of the Immortality Bus as the Entropy Bus, and of ourselves as trundling across Texas in a great mobile metaphor for the inevitable decline of all things, the disintegration of all systems over time.

The elements shall melt with fervent heat, the earth also and the works that are therein shall be burned up.

"Entropy sucks," said Roen.

"It is what it is," said Zoltan. "It absolutely is what it is."

I had begun, I realized, to feel some strange affinity with these two men: not through any deeply felt sympathy for their mystical aims, but rather through being with them, traveling with them, eating in the same truck stops and sleeping in the same motels and listening to the same endless loop of Tom Petty and the Heartbreakers on the ancient cassette deck of the Wanderlodge. It was a comradeship of sorts; we were confreres in futility, which was perhaps the best that could be said of any association of humans. But then they would never have consented to such a description of our situation, and so it was, in that sense, no kind of comradeship at all.

This question of futility was raised many times on the Wanderlodge. Zoltan and Roen believed that life was rendered meaningless by death. If in the end everything was lost, they asked, what was the point of anything?

I did not feel qualified to answer this question, but I tried to make a case for life as it currently stood, which meant trying to make a case for death. Wasn't it the fact that life ended, I asked, that gave it what meaning it had? Wasn't it the very fact that we were here for so brief a time, that we could be gone at any moment, that made life so intensely beautiful and terrifying and strange? (Then again, wasn't the idea of meaning itself an illusion, a necessary human fiction? If a finite existence was futile, wouldn't immortality be just a state of endless futility?)

There was no beauty in finitude, they said, no meaning to be extracted from oblivion. My arguments, Roen insisted, were transparently motivated by a "deathist" ideology–a need to protect myself against the terror of death by trying to convince myself that death was actually not so terrible. As crazy as most of what Roen had said sounded to me, he was, I thought, basically right about this. This was an idea that had been suggested to me, in one form or another, by many of the transhumanists I had spoken to over the previous eighteen months–by Natasha Vita-More, for instance, by Aubrey de Grey, by Randal Koene.

We drove through the emptiness. *Don't Mess with Texas.* Ruptured armadillos, rotting in the desert heat. *Stand with Israel.* Zoltan swigged at intervals from a magnum-sized vessel of greenish energy drink he'd picked up on our last Walmart stop. We talked for hours, and then for hours more we said nothing at all. We listened to the Tom Petty cassette straight through, twice, three times. "Running down a dream," he sang, "that never would come to me." Forty minutes later, he sang it again.

What was it that was going on here? The whole enterprise suddenly seemed to me like an absurdist parody of social privilege: three white men traveling across a wilderness, protesting against the final injustice they would one day suffer in common with all creatures, the great leveler that must itself be leveled. Was dying of old age not, in this sense, the ultimate First World Problem?

About an hour east of Ozona, we pulled off the Interstate onto a narrow side road so that Zoltan could remove the barbecue tray, which was by now almost overbrimming with leaked oil. We were on the perimeter of a vast ranch, a flat and half-barren landscape of scrub grass and squat cactuses as far as the human eye could see. I went behind the bus to take a piss, and as I did so I looked up and counted five vultures idling overhead, like predator drones in the inverted abyss of the sky. I tried to imagine how we might have appeared to the serenely primordial eyes of these eschatological beasts, three medium-sized mammals lumbering upright, without apparent purpose, around a great coffin-shaped Leviathan. But what could any of this—men, coffins, journeys—possibly mean to these creatures, to whom nothing was required to mean anything? Probably, we were irrelevant to their view of the landscape, being too large to kill, and not yet otherwise dead.

I struggled to remember a line from the eighth of Rilke's *Duino Elegies,* where he writes about the freedom in which animals live—"the Open" upon which they look out, and which vista is unavailable to us, oriented as we are always toward the overbearing presence of our own finitude. Back on the bus, I Googled it on my phone, and found it: "We, only, can see death; the free animal/has its decline in back of it, forever, and God in front, and when it moves, it moves/already in eternity, like a fountain."

Later, as we barreled along the Interstate, Roen cheerfully drew our attention toward a gigantic billboard that read "IF YOU DIE TODAY, WHERE WILL YOU SPEND ETERNITY?"

"In the ground," he said. "In the ground."

He told me about an accident he'd had when he was six. He'd fallen from his bike, and punctured his spleen and had very nearly died from internal bleeding. Weeks in the hospital, then a recovery; but a darkness had been revealed to him, a black terror beneath the thin surface of the world. Every night, he awoke gasping from the same nightmare, in which he had died in his sleep, in which he felt himself lying there in his bed feeling nothing, an impossible body. Every night this same experience of a thing that could not be experienced, this same vision of a thing that could never be glimpsed. This was the beginning of his move away from the religion of his parents, he said, this vision of the nothingness that awaited him after death.

Farther east, at a roadside rest stop, Roen switched on his video camera and approached two young women seated at a picnic area, beneath a corrugated iron shelter flanked on either side with giant wagon wheels. Pointing the camera in their faces, he asked them if they feared death. They seemed more bemused than intimidated, but I wanted no part of this exchange, and so I wandered off toward the other side of the rest stop. My path was intersected by two young men. They wanted to know why my friend was filming their girlfriends. I pointed to the Wanderlodge, and told them we were on the campaign trail with a third-party presidential candidate, and that Roen was making a documentary.

"That guy's running for president?" said the larger of the two, peering skeptically at Roen–Roen, with his Joan Baez hair, his knee-length shorts, his unblinking righteous eyes.

"Not that guy, the other guy," I said, indicating Zoltan, who was standing beside the Wanderlodge finishing up a phone call. "That's his campaign bus. I'll introduce you guys if you want."

We all–me, Roen, the two young women, their boyfriends–went over to Zoltan, who met these constituents with expansive gestures, warm greetings, statesmanlike handclasps.

"So what's up with the bus?" said the smaller and tougher-looking guy.

"We fixed it up to look like a giant coffin, to raise awareness about death."

"It doesn't look like a giant coffin," said the guy. "Looks more like a giant turd."

Zoltan tactfully ignored this remark, and explained, somewhat superciliously, that the goal of the campaign was "to promote investment in longevity science so that you can live a longer life."

A sturdy, low-slung man who looked to be in his mid-thirties descended from the cab of an adjacent truck, stretched briefly, and narrowed his eyes at the Wanderlodge, the assembled group, before ambling over. He was wearing purple basketball shorts, a voluminous black T-shirt, Oakley shades. His name was Shane, he said, and he was driving cross-country toward Florida.

"You got some kind of political thing going on here?" asked Shane.

"Yes we do," said Roen. "Do you want to live forever?"

"Sure I would," said Shane. "I'm scared as hell to die. Who wouldn't want to live forever?"

"What we're trying to do," said Zoltan, "we're trying to promote using science to end aging and death. We work with some scientists who are really close to stopping the progress of aging. It sounds crazy, I know, but it's true. I'm actually one of the leading third party candidates in America. Our party is called the Transhumanist Party."

"What does 'transhumanist' mean?" asked Shane.

"Well, it means a lot of things. Not dying is one of them. A lot of us want to evolve into machines. My father, for instance, had four heart attacks in quick succession quite recently. That sort of thing wouldn't happen to people if we were machines."

"Right on," said Shane, politely. "I could get behind that."

He talked and listened a little longer, then excused himself to continue his journey eastward. He couldn't linger too long at any rest area, he explained, because his progress and speed were closely monitored by an onboard computer in his vehicle, which reported back to his employers, alerting them if he had stopped for longer than the permitted time, or if he was going faster than the permitted speed to make up for those lost minutes and hours. I thought for a moment that he might have been making a subtle point about the extent to which capitalism had already evolved many of us into machines, or even alluding to an imminent future in which his aforementioned employers would replace him with self-driving technology, but as he hauled himself into the cab of his truck and waved back at us, I decided that he had probably not been making any such finely insidious point. He seemed like more of a straight-shooting type of guy.

"What do you say," asked the newsman, "to people who accuse you of trying to play God?"

We were standing in a lavishly tree-lined street in the upscale residential neighborhood where the campaign event was about to take place, and Zoltan was being interviewed for Austin TV news. He was dressed in a shirt and slacks, his hair combed meticulously back from his high-domed forehead.

"I would agree that we are, in fact, trying to play God," he said.

It was to me that he said this—or at any rate it was at me he was looking when he said it. The bearded and lavishly perspiring cameraman, who was double-jobbing as a reporter, had requested that I stand to one side of him as he did so; Zoltan would, in this way, appear to be addressing a dedicated news reporter, rather than a guy who, presumably because of budget cuts, had been forced to do two jobs simultaneously.

And so although it was me he was looking at, he was, more properly, speaking to the television viewers of Austin and, beyond that, to the people of the Internet, the unseen demos of clicks and engagements. The experience was vaguely uncanny, as though I myself had ceased to exist, had dissolved into a vacancy through which the world itself could be addressed.

This was a thing that had been happening to me lately. I had started to see myself as a mechanism through which signals were passed. I would be sitting on the bus, jotting down snatches of conversation in my notebook, details of scenery or sensation, and I would see myself as a primitive device, a machine for the recording and processing of information. I would be at the checkout in a cavernous Walmart, paying for snacks, and I would see myself as one of many millions of mechanisms in a vast and mysterious system for the upward transfer of wealth. I knew, of course, that this was the result of my overexposure to mechanistic ideas, but on some level I recognized that I had always seen myself in this way. Nothing is stranger to man, as Čapek put it, than his own image. Nothing is stranger than that which is most familiar.

"And what made you decide to run for president?" asked the cameraman who was also a reporter.

"I believe," said Zoltan, "that we should take technology as far as it can take us." His hand gestures had the practiced decisiveness of a real-deal politician; in the presence of the camera, as he gazed unblinkingly into my eyes, he had taken on a plausibly presidential aura; he seemed, suddenly, a vast physical presence, a great hollow monument to his own significance.

"And that includes," he said, "becoming technology ourselves. At some point, we are going to become more machines than human beings. That's what my presidential campaign is advocating for. That's the conversation I'm trying to start."

A cluster of young men approached us. They were part of the Austin biohacker group, and they were here for the campaign event. They had names like Alec, Avery, and Shawn; they were, for transhumanists, a startlingly frat-brovian contingent, all laid-back Texan vibes, loose-fitting vests, hypertrophied upper bodies.

Roen received them in his usual manner, by avoiding traditional greetings in favor of immediate interrogation as to their stance on eternal life.

"I'm down," said the guy called Alec, as though Roen had just asked him if he wanted to go in on an ounce of weed. "Let's do it. Let's make it happen. Life is awesome."

"Right?" said Roen. He glanced meaningfully at me, a look that I took as mild admonishment for previous exchanges in which I'd expressed reservations about absolute judgments on the awesomeness of life.

"Got a lot to accomplish, man," said Alec. "I can't be dying at eighty. I need at *least* two hundred years to get my shit done. Maybe two-fifty."

"Right? I mean, when you see a really old person, what do you think?"

"I think it sucks, is what I think," said Alec. "I think that can't be very enjoyable."

We went through into the house where the event was to take place, a small split-level open-plan that was almost entirely devoid of furniture. I was given to understand that the place was shared by a loose cohort of biohackers; it was not apparent who did and did not live there, but it seemed to be a kind of transhumanist commune, or futurist frat house. Even for an event of this sort, the gathering was overwhelmingly male.

As we entered the sunken living area, we squeezed past a tall and powerfully built man wearing a baseball cap and tightly fitted T-shirt. He was swigging a beer, talking to a smaller man with pink-striped hair and multiple facial piercings. The tall dude had a lazy drawl, and was leaning against the doorframe with the informal facility of a ranch hand taking his ease against a fence.

"Man," he was saying, "the dude was just *real* stoked about the code, so I totally let him into the GitHub."

A long-haired young guy wearing an intricately embroidered Indian-style shirt introduced himself as the organizer of the Biohack Austin group. His name, he said, was Machiavelli Davis, though we were warmly encouraged to call him Mac. He was originally from Singapore, and he was a biology graduate student at the University of Texas.

While Zoltan went through the particulars of the evening's speeches with him, I wandered over to a table where a man in flip-flops and a T-shirt with a cartoon of a beer wearing sunglasses was futzing with a complicated-looking device. The device was composed of a small aluminum suitcase, and featured many wires and electromagnetic relays, along with amorphous lumps of magnesium and plastic cups of water.

The man, whose name was Jason, told me that the whole setup was a prototype of a product he was developing called Heliopatch, which he described as a "functional life extension pod." In combination with the user's body, he said, the device functioned as a battery, with the magnesium patch acting as an anode and the user's body as a cathode. When the patch was applied, the magnesium corroded, he said, releasing electrons and positive ions into the body, thereby neutralizing the free radicals that caused cell damage, and reducing the aging process. A while back, he told me, he'd implanted a small magnesium patch into the inside of his left cheek, and kept it there for a month. He then asked a bunch of friends which side of his head had less gray hair. "They were all like, definitely the left side," he said. "Every single one of them."

The living room was getting crowded now, and Machiavelli had started giving a speech. He was saying something I didn't quite catch about having spent some months in a Buddhist monastery in Thailand. Then he said something else about how the time in which we were living would see one of the most massive changes in human history; everything, he said, was "set up and ready to fall." The rise of the biohacking movement, he said, of people's ability to edit genes and augment their bodies, would be the defining influence on this generation, and the generations to come. A couple of weeks from now, he said, he was organizing a trip out to the desert with the Biohack Austin group. The plan was for everyone to take a vision-enhancing eyedrop solution—a special formula made from a molecule called chlorin E6, found in the eyes of certain deep-sea fish, which amplified photonic signals to the brain by a factor of two—and to gaze at the light of the stars with superhuman sight. This experiment, he said, had been successfully conducted with rats, and he and his fellow biohackers would be the first humans to experience it.

"What humanity does," he said, "is experiment with itself. It's something we have a birthright to do. That, to me, is what freedom means: to practice liberty with your own body and mind."

Zoltan picked up this thread in his own fluent and apparently unscripted speech. History was being created, he said, with this movement, with this campaign, which was not about getting votes, but about raising awareness of the coming Singularity, and the importance of living long enough to experience it. He believed, he said, in morphological freedom—in the absolute and inalienable right of people to do whatever they wanted with their bodies, to become more than human.

"I look forward," he said, "to the day when we can use technology to make ourselves more machinelike."

We hung around for an hour or so afterward, and Zoltan spoke to some people who were making a documentary about transhumanism, and a woman from a magazine who'd come to interview him. At some point Roen gave an impromptu speech of his own. He delivered this oration "in character" as a generic hipster, wearing a pair of black-framed glasses, and a slightly strained knowing smirk on his face. This was a role he'd been tinkering with in the videos he'd uploaded to the Eternal Life Fan Club's Facebook page over the course of the campaign.

"You guys aren't mainstream," he said to the assembled biohackers, most of whom seemed mildly perplexed by the performance. "You still have your childlike imaginations. If you want to take your non-mainstream-ness to the next level, you're gonna have to live forever. You know what the most mainstream thing ever is? Dying. Dying is totally mainstream. Being dead in the ground is totally mainstream. Vote for Zoltan if you want to live forever!"

I'd seen this performance of Roen's before, and had advised him that it was a little too broad in its depiction of the hipster trope—that it seemed like more of a caricature of a caricature than a representation of any actual person, and that, furthermore, the injection of performative irony into his delivery tended to obscure the absolute earnestness of his message. But right now, perhaps because of the unusually strong homebrew beer I was drinking, I was enjoying it immensely, and I felt a strange tenderness for him swelling in my chest, an almost fraternal instinct of protection, very much at odds with any properly journalistic imperatives.

I had agreed with practically nothing that had come out of his mouth in the entire time we'd spent together. He was as strange a person as I had ever met, and I had met a great many strange people over the past year and a half. I found myself hoping that he would not be disillusioned, that he would maintain, as long as he lived, the sense of his own exemption from death. His very belief that existence was rendered meaningless by death was, I thought, precisely what seemed to afford his life a sense of purpose, a sense of direction. This, in the end, was why humans would always look for meaning, and would always find it in some or other variety of religion. You do what you can with the strangeness of being here, for the time being.

As soon as the media people had moved on, Zoltan wanted to hit the road; the party was still building momentum, but he had an early flight the next morning to Miami, where he had a corporate speaking engagement, and he needed to pilot the Wanderlodge to a place across town where he'd arranged, through the good offices of Machiavelli, to park it until the next leg of the tour. And so he completed a valedictory round of handshakes, and then we boarded the Immortality Bus once more.

An hour or so later, we were in the backyard of an empty house on the far outskirts of the city, waiting for a cab to come and take us to our respective hotels. Zoltan and I were drinking the last of the Immortality Bus's booze stash, a bracingly potent vodka whose bottle featured a flashing digital display, a *Jetsons*-like vision of the future of vodka bottles. I was feeling a little light-headed from the booze, and from the small amount of weed I'd smoked at the party before remembering I hated smoking weed, and so I climbed down into the yard to take the air. The night was warm and fragrant, and alive with the gentle chirping of crickets. I stared up at the stars, feeling pleasantly out of it. It was good to be outside, to be in the world, to be a living animal.

The more I listened, the more urgent the chirping of the crickets seemed to become. I remembered then having read something in the news a couple of weeks previously about a cricket infestation in the plains of the southwestern states that had been particularly intense in the area around Austin. The swelling of the insects' numbers had to do with the summer having been unusually cool, and unusually wet. Crickets were impelled to mate, apparently, by a cooling of the air, which forewarned them, on some primordial level, of their own impending death. The chirping I was listening to, I now realized, was the sound of thousands of male animals expressing their urge to reproduce, in the instinctive knowledge of their own approaching demise. The sound seemed to be intensifying, and to be coming at once from everywhere and nowhere, to be generated by the night itself.

I heard the chirp of Zoltan's phone from across the yard. Our cabdriver calling, probably. I breathed in deeply, assimilating the warm and complicated air, the fragrant night. In my tipsy state, it seemed outright implausible that all of this would one day be beyond my reach, that one day I would die and never again breathe this air, or hear these sounds—crickets, traffic, words, vibrating phones: the interwoven signals of animals and machines—or feel the hopeful surge of alcohol in my blood, the world advancing its uncertain promise. It seemed ludicrous to think that this was it: just this once, and never again.

I heard the hollow slam of the Immortality Bus's door, and Zoltan calling my name. Our cab had pulled up at the curb. I took a last look at the looming apparition of the bus, the great brown sarcophagus of the American highway, and was momentarily taken by the facile charm of its standing as a metaphor for life itself: an incomprehensible and futile journey, in a vast coffin-shaped recreational vehicle, out of one nowhere into some other. I walked toward the street, toward

Zoltan and Roen, deciding to tell them about this life-as-coffin-bus idea, and to tell them that I was glad to have been on this journey for a while, whatever it meant or did not mean. But by the time I got to the car and slid in beside Roen, Zoltan was already sitting up front, passionately laying out the coordinates of the posthuman future to our cabdriver, and the moment had passed.

A Short Note on Endings and Beginnings

IT SO HAPPENED, not long after my time among the transhumanists had come to an end, that I was lying on a hospital gurney and gazing at a large computer screen on which the interior of my body was displayed. I was looking, specifically, at the fleshy lining of my colon, and I was pleased, in a detached sort of way, to note its cleanliness. The twenty-four hours I had endured without food, and the barbarously effective series of laxatives I'd been prescribed, had established my interior surfaces as ready for prime time. I was in a position to note these things with detachment, rather than horror, because I had just been given a dose of an extremely potent synthetic opiate.

"Could you turn on your side please? Toward the screen, yes. And pull your knees up toward your chest. There we go."

I had been told that this dosage would cause me to sleep throughout the colonoscopy, but this was not the case. I felt I could have slept had I wanted to, had I just closed my eyes and allowed myself to drift, but I found that I was not unhappy to be awake. I was looking at the inside of myself on a screen, and I was feeling at peace for the first time in weeks–for the first time since seeing the blood in the toilet bowl; since hearing my doctor say I needed a colonoscopy; since being confronted with the possibility of bowel cancer–the possibility that, far from being midway on the journey of my life, I might in fact be nearing its end.

It had been a dark and constricted time: a time of dawn awakenings, dreams of suffocation. A time of mortal disquiet in the bathroom, blood on white ceramic. It had been a time of reaching to switch off the car radio during ads for life insurance, of my wife and me smiling less indulgently at our son's persistent questions about death.

I felt no increased attraction toward cryonic suspension, or whole brain emulation, or radical life extension; I felt no greater urge toward becoming a machine. But I was by no means unflinching, either, in the face of my own animal mortality. I flinched constantly. I flinched as though my life depended on it. I felt a good deal less sanguine about this matter of my own mortality than I had on the Immortality Bus. Roen had been right, of course: I was guilty of deathism.

But on the gurney, at a palliative remove from the whole business of myself, all of that was an abstraction. I was a physical body, regarding itself on a screen, and I was also not a body at all, but a consciousness, or a sensation of consciousness. On the screen, a hooked metal instrument appeared, a small and intimately malevolent thing inside the body that I understood to be mine. A tiny movement, a tearing of flesh. A little blood, a withdrawal. This, I understood, was the biopsy.

The phrase "meat machine" was offered to me from nowhere, as if for inspection. I didn't so much think about it as hold it for a moment, before letting it go.

I considered, with detachment, the detachment of my own considerations. I was thinking clearly for the first time, though what I was doing was scarcely thinking at all. I was, finally and literally, up my own arse. I had chosen the sedation because I feared the discomfort of penetration, but I was glad to be awake now, to be witnessing this merger of self and technology, this dissolution of boundaries. The paradoxical effect of this infiltration was that I felt inviolable, as though nothing could touch me. I felt that I finally understood what it might mean to be posthuman. In retrospect it was obviously the drugs, but at the time it felt like the technology.

A few minutes or a few hours later—it was impossible to tell, and irrelevant anyway—the gastroenterologist who'd performed the procedure appeared by my side. I was elsewhere now, back in the room where they'd inserted the cannula in the crook of my arm, though I had no memory of being taken there. It was strange, he said, a strange inflammation, but nothing sinister. Diverticular colitis, most likely. Not cancer, then? No, not cancer.

He said some more things, the general thrust of which was that I was not dying—not, at least, in any immediate sense—and then he went away.

I closed my eyes, and saw again the screen, the inner spaces, the soft and clean interior of the body. Painlessly, the opiate veil was being withdrawn. For a moment there, I had been outside of myself, outside of time. For a moment there, I had been at one with the technology.

I lay on my back on the gurney, and I looked at the cannula in my arm, one of the two channels through which science had recently entered my body. I slowly clenched and unclenched my hand, listening to the soft click of bones and ligaments in the wrist, its arcane technology of flexion and torsion. I thought of a question my son had put to my wife and me some days previously, while looking at his own hand.

"Why do we have *skin?*" he had asked, as though coming to sudden awareness of some long-standing absurdity.

"So that our skeletons are covered up," my wife had replied.

I turned to my side and closed my eyes, and felt a gentle surge of relief at the realization that whatever was going on inside me was not going to kill me—that my skeleton would remain covered up for the foreseeable future, and that the machinery, the substrate, would continue to function, if perhaps a little less efficiently from this point on. I felt the distinction between myself and my body dissolve, like a dream of some impossible suspension. I was coming back to myself, whatever that meant. The problem of death—for this particular animal, in this particular instance—had been solved.

—

At time of writing, Zoltan's campaign is still a going concern. Roen is still filming, still asking people whether they want to live forever, and if not, why not.

At time of writing, no minds have been uploaded, no patients awakened from cryonic suspension and returned to life. No artificial intelligence explosion has taken place, no Technological Singularity.

At time of writing, I regret to say, we are all of us still going to die.

Among the transhumanists, among their ideas and their fears and their desires, I sometimes found myself thinking that the future, if it came, would vindicate them by forgetting them. I sometimes found myself thinking that the situation of our species might change so totally in the decades and centuries to come that it would no longer be necessary to speak of a merger of humans and technology—that it might, in other words, no longer make sense to speak as though such a distinction existed. And that transhumanists, if they were to be remembered at all, would be thought of as a historical curiosity, as a group of people who spoke, out of their time, in their feverish way, of what was in fact to come.

I could tell you that I have seen this future, that I bear news of some great convergence or dissolution that awaits us. But it is only true, in the end, to say that I have seen the present, and the present is strange enough to be getting along with: filled with strange people, strange ideas, strange machines. And even this present is not knowable, or graspable—but it can at least be witnessed, glimpsed in brief flashes, before it's gone. And it's a futuristic place, the present, very much like the past. Or at least it was at the time I encountered it, which is already receding into oblivion, into memory.

What I came to feel, in the end, is that there is no such thing as the future, or that it exists as a hallucinatory likeness of the present, a comforting fairy tale or a terrifying horror story that we tell ourselves in order to justify or condemn the world we currently live in, the world that has been made around us—out of our desires, in spite of our better judgment.

I am not now, nor have I ever been, a transhumanist. I am certain I would not want to live in their future. But I am not always certain I don't live in their present.

What I mean to say is that I am part machine: encoded in the world, encrypted in its strange and irresistible signals. I look at my hands as they type, their hardware of bone and flesh, and I look at the images of these words as they appear on a screen, my screen: a feedback loop of input and output, an algorithmic pattern of signal and transmission. The data, the code, the communication.

I am remembering, now, a question that Marlo Webber asked me on my last night in Pittsburgh, down in the basement, with its mingled aromas of caramel vape smoke and sweat and burnt silicon.

He said: "What if we're already living in the Singularity?" And as he said this, I remember, he picked up his smartphone and weighed it in his hand with rhetorical intent, flipping it, catching it. He was talking about the phone, I knew, but also all that it was connected to—the machines, the systems, the information. The unknowable vastness of the human world.

"What if it's already started?" he said.

It was a good question, I told him. I'd have to think about it.

Acknowledgments

I could not have started this book, let alone completed it, without the support and encouragement of my wife, Amy. My gratitude for her love and wisdom exceeds any capacity I possess to express it in words. From the very beginning, the unseen hand behind this project has been that of my agent, Amelia "Molly" Atlas. I'm extremely fortunate to have her, and the good people of ICM, looking out for me. My deepest thanks are also due to Karolina Sutton at Curtis Brown in London, and to Roxane Edouard. Yaniv Soha at Doubleday has been a wise and sustaining presence over the course of my writing this book. His enthusiasm and subtle editorial guidance have been invaluable. My thanks also to Margo Shickmanter for her work on the project. From the beginning, Max Porter at Granta has been an abundant source of insight and encouragement, and a generally terrific person to have in my corner.

For their various hospitalities, kindnesses, and acts of professional and personal decency, I am also forever grateful to the following people: my parents, Michael and Deirdre O'Connell; Kathleen and Elizabeth Sheehan; Susan Smith; Colm and Alexa Bodkin; Lydia Kiesling; Dylan Collins; Ronan Perceval; Mike Freeman; Sam Bungey; Yousef Eldin; Daniel Caffrey; Paul Murray; Jonathan Dykes; Lisa Coen; Katie Raissian; Chris Russell; Michelle Dean; Sam Anderson; Dan Kois; Nicholson Baker; Brendan Barrington, and C. Max Magee.

This book could not have been written without the cooperation and assistance of the following people: Zoltan Istvan, Roen Horn, Max More, Natasha Vita-More, Anders Sandberg, Nick Bostrom, David Wood, Hank Pellissier, Maria Konovalenko, Laura Deming, Aubrey de Grey, Mike La Torra, Randal Koene, Todd Huffman, Miguel Nicolelis, Edward Boyden, Nate Soares, David Deutsch, Viktoriya Krakovna, Janos Kramar, Stuart Russell, Tim Cannon, Marlo Webber, Ryan O'Shea, Shawn Sarver, Danielle Greaves, Justin Worst, and Olivia Webb.

A Partial List of Works Consulted

Adorno, Theodor W., and Max Horkheimer. *Dialectic of Enlightenment: Philosophical Fragments*. Stanford: Stanford University Press, 2002.

Arendt, Hannah. *The Human Condition*. Chicago: University of Chicago Press, 1989.

Armstrong, Stuart. *Smarter than Us: The Rise of Machine Intelligence*. Berkeley: MIRI, 2014.

Barrow, John D., and Frank J. Tipler. *The Anthropic Cosmological Principle*. Oxford: Oxford University Press, 1986.

Becker, Ernest. *The Denial of Death*. New York: Free Press, 1973.

Blackford, Russell, and Damien Broderick. *Intelligence Unbound: The Future of Uploaded and Machine Minds*. Chichester: John Wiley & Sons, 2014.

Bostrom, Nick. *Superintelligence: Paths, Dangers, Strategies*. Oxford: Oxford University Press, 2014.

Čapek, Karel. *R.U.R. (Rossum's Universal Robots): A Fantastic Melodrama*. Trans. Claudia Novack. London: Penguin, 2004.

Chamayou, Grégoire. *Drone Theory*. London: Penguin, 2015.

Cicurel, Ronald, and Miguel Nicolelis. *The Relativistic Brain: How It Works and Why It Cannot Be Simulated by a Turing Machine*. Montreux: Kios Press, 2015.

Clarke, Arthur C. *The City and the Stars*. New York: Harcourt, Brace, 1956.

Descartes, René. *Discourse on Method and Meditations on First Philosophy*. Trans. Donald A. Cress. Indianapolis: Hackett Classics, 1998.

——. *Treatise of Man*. Trans. Thomas Steele Hall. Amherst, NY: Prometheus, 2003.

Dick, Philip K. *Do Androids Dream of Electric Sheep?* New York: Doubleday, 1968.

Dyson, George. *Darwin Among the Machines: The Evolution of Global Intelligence*. London: Penguin, 1999.

Ellis, Warren. *Doktor Sleepless*. Rantoul, IL: Avatar Press, 2008.

Emerson, Ralph Waldo. *Nature and Selected Essays*. New York: Penguin, 2003.

Esfandiary, F. M. *Up-wingers*. New York: John Day, 1973.

Ettinger, Robert C. W. *The Prospect of Immortality*. Garden City, NY: Doubleday, 1964.

Foucault, Michel. *The Order of Things: An Archaeology of the Human Sciences*. New York: Pantheon, 1971.

Gibson, William. *Neuromancer*. New York: Ace, 1984.

Gray, John. *The Soul of the Marionette: A Short Inquiry into Human Freedom*. London: Penguin, 2015.

——. *Straw Dogs: Thoughts on Humans and Other Animals*. London: Granta, 2002.

Habermas, Jürgen. *The Future of Human Nature*. Cambridge: Polity Press, 2003.

Haraway, Donna. *Simians, Cyborgs and Women: The Reinvention of Nature*. New York: Routledge, 1991.

Hayles, Katherine. *How We Became Posthuman: Virtual Bodies in Cybernetics, Literature, and Informatics*. Chicago: University of Chicago Press, 1999.

Hobbes, Thomas. *Leviathan*. Cambridge: Cambridge University Press, 1991.

Jacobsen, Annie. *The Pentagon's Brain: An Uncensored History of DARPA, America's Top-Secret Military Research Agency*. New York: Little, Brown, 2015.

Jennings, Humphrey, Mary-Lou Jennings, and Charles Madge. *Pandaemonium: The Coming of the Machine as Seen by Contemporary Observers, 1660–1886*. New York: Free Press, 1985.

Kurzweil, Ray. *The Singularity Is Near: When Humans Transcend Biology*. New York: Viking, 2005.

Lem, Stanislaw. *Summa Technologiae*. Trans. Joanna Zylinska. Minneapolis: University of Minnesota Press, 2013.

Ligotti, Thomas. *The Conspiracy Against the Human Race: A Contrivance of Horror*. New York: Hippocampus Press, 2012.

Midgley, Mary. *The Myths We Live By*. London: Routledge, 2003.

——. *Science as Salvation: A Modern Myth and Its Meaning*. London: Routledge, 1992.

Moravec, Hans P. *Mind Children: The Future of Robot and Human Intelligence*. Cambridge, MA: Harvard University Press, 1988.

——. *Robot: Mere Machine to Transcendent Mind*. New York: Oxford University Press, 1999.

More, Max, and Natasha Vita-More, eds. *The Transhumanist Reader: Classical and Contemporary Essays on the Science, Technology, and Philosophy of the Human Future*. West Sussex: Wiley-Blackwell, 2013.

Noble, David F. *The Religion of Technology: The Divinity of Man and the Spirit of Invention*. New York: Alfred A. Knopf, 1997.

Pagels, Elaine. *The Gnostic Gospels*. New York: Random House, 1979.

Rothblatt, Martine. *Virtually Human: The Promise and the Peril of Digital Immortality*. New York: St. Martin's, 2014.

Searle, John. *Minds, Brains and Science: The 1984 Reith Lectures*. London: Penguin, 1989.

Seung, Sebastian. *Connectome: How the Brain's Wiring Makes Us Who We Are*. London: Penguin, 2013.

Shanahan, Murray. *The Technological Singularity*. Cambridge, MA: MIT Press, 2015.

Shelley, Mary. *Frankenstein*. London: Penguin, 2007.

Solnit, Rebecca. *The Encyclopedia of Trouble and Spaciousness*. San Antonio: Trinity University Press, 2014.

Teilhard de Chardin, Pierre. *The Phenomenon of Man*. New York: Harper Perennial, 2008.

Wiener, Norbert. *Cybernetics; Or, Control and Communication in the Animal and the Machine*. Cambridge, MA: MIT Press, 1961.

——. *The Human Use of Human Beings: Cybernetics and Society*. Boston: Da Capo, 1954.

Printed in the United States
by Baker & Taylor Publisher Services